EXPLORING
THE DEEP

INTERNATIONAL LIBRARY

ANDREI AKSYONOV AND ALEXANDER CHERNOV

EXPLORING THE DEEP

COLLINS · PUBLISHERS

London · Glasgow ·

FRANKLIN WATTS, INC.

New York

First edition 1979 (U.K.)
First published in the United States
of America by Franklin Watts Inc. in 1979

ISBN 0 00 100181–7 *(Collins)*
SBN 531 02126–2 *(Watts)*

CONTENTS

CHAPTER 1

WHY
WE STUDY
THE OCEANS

The whole history of human society is closely associated with the sea, not only with navigation but with the utilization of the sea's raw materials, its huge resources of food and, more recently, its fuel and energy. Jacques-Yves Cousteau, the well known explorer of the ocean depths, has predicted that the future existence of mankind will depend upon the sea. Geologist V. Smirnov and oceanographer L. Brekhovskikh, both members of the USSR Academy of Sciences, have also expressed the view that the threat of a mineral famine, in the literal sense, will compel man to turn to the oceans and begin their exploitation more actively.

What is known about the oceans

The International Geophysical Year began in July 1957. For the first time in history, scientists of 68 countries began simultaneous observations of various natural phenomena. The French and Swiss observed glacial movement in the Alps; while Soviet scientists studied the glaciers of the Caucasus, Pamirs and the Tien Shan ranges. Meteorological stations all over the world closely watched changes in the weather and submitted immediate results of their observations to specially equipped international centres. The flow of rivers, earthquakes, volcanic eruptions and other phenomena were also investigated.

This tremendous undertaking had a single aim: to better our understanding of all that occurs on our planet and in its atmosphere. For the earth is the common home of all mankind, and the more man knows about it the more wisely he will employ its natural resources. The International Geophysical Year was a triumph of co-operation by scientists of the whole world. Among those participating in its extensive work were those specialists who study the oceans—oceanographers and marine biologists.

The surface of the earth occupied by dry land has been plotted long ago in great detail on geographical maps. But, until recently, very little was known of the features of the immense ocean floor. It may be common knowledge that the peak of Mount Everest is the highest in the world but only lately have we determined the greatest depth in the ocean.

We have a sufficiently detailed knowledge of the winds blowing over land. Air currents are observed regularly by thousands of meteorological stations, whereas investigations of the atmosphere over the oceans and of ocean currents themselves are conducted only from time to time by research ships and from

The Perry Cubmarine, an American submarine designed for biological research on the continental shelf. Its body of plastic and glass is light and streamlined. The crew of two operate from a steel and aluminium compartment in which there are 12 large portholes, giving good all-round vision.

The Soviet research vessel Vityaz *has explored the oceans for many years. During more than 70 expeditions major scientific discoveries have been made in the fields of ocean biology, the geology of the sea bed and the movement of ocean currents.*

certain islands. The quantity of rain and snow falling on land is well known and has been registered for many years, but scientists have little idea of the amounts falling on the sea, which represents 71 per cent of the earth's surface.

The International Geophysical Year began an era of major and minor oceanographic discoveries. Scientific expeditions from various countries were despatched to different regions of the oceans. One such expedition of special significance was that of the Soviet research ship *Vityaz* which sailed in 1959 to the region of the Mariana Trench in the Pacific. This resulted in the discovery by echo-sounder of the deepest part of the ocean, 11,022 metres, which was later named Vityaz Deep.

In October 1959 scientists from countries all over the world held the First International Oceanographic Congress in New York where many scientific papers were read, some concerning the very latest discoveries. General conclusions were reached which assessed man's present knowledge of the oceans. In 1966 the second such congress was held in Moscow. It was attended by over 2,000 people, a substantial increase over the 600 participants of the New York congress. In their papers, the scientists reported a great number of discoveries that had been made in the last few years, and new maps of the ocean floor were on display. The Moscow congress led to the organization of several long-term expeditions.

One international expedition is studying the permanent current in the western Pacific—the Kuroshio, or Japanese current, which carries its warm waters, heated in the tropics, far to the north to the shores of Japan and the Sakhalin and Kurile Islands. From year to year the Japanese current varies; it is sometimes strong and sometimes weak. It affects the weather, ice formation and, what is most important, the abundance of food-fish, such as herring, in the northern part of the Pacific.

Scientists from several countries

on board the American ship *Glomar Challenger* have been engaged in deep sea drilling projects for a number of years. Studies of the Mediterranean and Caribbean seas are also being conducted on the basis of international programmes.

Another international expedition spent several years in the Indian Ocean. Today, this once mysterious body of water is as well known as the Atlantic. A geological and geophysical atlas of the Indian Ocean was published in the Soviet Union in 1975. It contains many new and detailed maps of the ocean floor. Soviet oceanographers investigated the relief of the ocean floor and the distribution of deposits, as well as the deep-lying structure of the earth's crust, submarine earthquakes and magnetic irregularities.

In our time the oceans are being studied in many ways. The best-known and most widespread method is the use of special research ships. Dozens of such expeditionary vessels depart on well-charted routes following definite, detailed programmes. They plot and measure ocean currents, study the distribution of plankton (a term for microscopic animal and plant life passively carried by the motion of the water) and small animals used as food by fish. They estimate the abundance of food-fish and make detailed explorations of regions of the sea bed which are rich in useful minerals.

In the last few years, man has learned to study the sea from space vehicles. Cloud formations above the oceans are clearly visible from space, as are atmospheric whirlpools or cyclones. From orbit, the water temperature at the ocean surface can be measured, wave phenomena observed, and the distribution and density of the drift ice determined. These measurements and observations from space provide useful information about a vast area of the oceans, but in accuracy and detail they are inferior to the information obtained in marine expeditions.

Finally, the oceans can be studied from the "inside". This is done by scientists wearing aqualungs or using special research submarines to work underwater. These methods yield unsurpassed results both in completeness and accuracy.

All the questions posed by the oceans can be divided into three groups. Each is concerned with the practical interests of mankind. First is the need for a precise knowledge of ocean currents, waves, tides and ice conditions in order to provide

The buoy-laboratory of Monaco's oceanographical museum is an inhabited steel island in the sea. The buoy's body, much of it underwater, is as high as a 20-storeyed building. Here scientists from the museum are engaged in daily research.

Marine life on the sea bed is rich and diverse. These sea pens look like graceful plants but are in fact animals and belong to the group called coelenterates.

In the Far East some species of sea cucumber are caught for food. The body of the animal is dried and used to make soup.

A characteristic feature of the young angelfish is that it differs greatly from its parents in both colour and shape.

The movement of water has sculpted this piece of crumbled basalt lava so that it looks remarkably like the head of a seal.

essential information for successful navigation.

Secondly, data on the distribution of life in the oceans and the relationship between the various groups of animals and plants are required for the proper management of the fishing industry. It is necessary to know where, when and how much fish can be caught without danger to natural reproduction.

And lastly, the structure of the earth's crust, the occurrence of certain deposits and the location of useful minerals are factors that must be made clear before these minerals can be extracted from the ocean bed.

The ocean and the atmosphere

For several years now, scientists of many countries have been working on an international project called GARP (Global Atmospheric Research Project). The aim of the project is to find some reliable method of forecasting the weather. It is hardly necessary to stress how important it is to know in advance of forthcoming droughts, floods, cloudbursts, disastrous winds, tropically generated storms or spells of extremely hot or cold weather. So far, no one can make such forecasts with accuracy. The main reason for this is lack of adequate information on the state of the atmosphere over the sea.

Almost all the water falling on land in the form of rain or snow is drawn up into the atmosphere from the surface of the world's oceans. In the tropics the waters of the oceans are heated to a high temperature and currents carry this heat to high altitudes. As a result, immense whirling cyclones are formed above the oceans which influence the weather on land.

The world's oceans are an area where weather is generated. Scientists are trying to develop a mathematical model of the interaction between the oceans and the atmosphere and several years of very accurate and continuous meteorological observations from ships, airplanes and weather satellites have been made.

Also of primary importance is a knowledge of the entire system of ocean currents. Besides heat and cold, currents carry nutrients such as mineral salts which are required to sustain plant and animal life in have been made that revealed how inaccurate these maps were and how complicated is the whole picture of water circulation in the oceans.

There is a permanent current at the surface of the Pacific Ocean flowing from east to west. It is driven by strong, regular winds, the trade winds. These winds and the current itself have been common knowledge for a long time, since the days of sailing ships. But now, for the first time, another current has

the depths of the sea. Sailors began to collect information on currents as far back as the 15th and 16th centuries, when sailing ships began to venture into the open sea. In our time, all navigators employ detailed maps of surface currents. In the last 20 or 30 years, however, discoveries been discovered at a depth of from 300 to 500 metres that flows from west to east. This countercurrent was named the Cromwell Current. Another powerful deep current, named the Lomonosov Current, was discovered, explored, measured and mapped in the equatorial zone

There is intense volcanic activity in the ocean depths and here an underwater volcano is seen erupting.

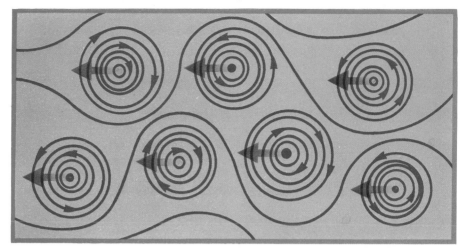

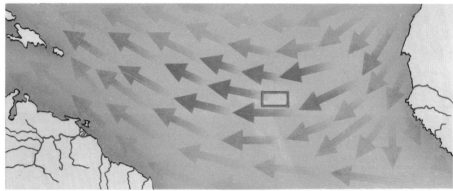

Scientists recently made the discovery that vast whirlpools move about the Atlantic and Pacific Oceans at tremendous speeds. They were detected in the area of the northern trade wind currents in the Atlantic Ocean. The rectangle in the lower picture indicates the place where research was conducted under the Polygon-70 experiment.

Sea lilies, among the oldest forms of present-day echinoderms, attach themselves permanently to the sea bed. Specimens have been found 3,000 metres down.

of the Atlantic. Similarly, the Antillean-Guiana deep countercurrent was discovered in the western part of the Atlantic Ocean, and a counter-Gulf Stream was found.

In 1970 Soviet oceanographers conducted a highly interesting investigation. A series of buoy stations was set down in the tropical zone of the Atlantic. Each station continuously registered the currents at various depths. When the data had been processed and examined, an important fact came to light: the existing concept of the comparatively uniform permanent trade wind current, driven by the northern trade winds, was not correct.

It was found that vast whirlpools, dozens or even hundreds of kilometres in diameter, travel along the zone of the trade wind current. The

centre of such an eddy travels at a speed of ten centimetres per second, but at the periphery of the whirlpool the currents have considerably higher velocity. This discovery made by Soviet scientists was later confirmed by American investigators and in 1973 similar eddies were traced by Soviet expeditions in the northern Pacific. This experiment, known as Polygon-70, yielded such promising results that at the present time an international expedition is being prepared for investigating this phenomenon on a wider scale.

Special probes were developed and used for the first time in 1969 to measure continuously the temperature and salinity of the water at various depths in the ocean. Previously, the temperature was measured with mercury thermometers at several points at different depths where samples of the water were taken with bathymeters. But such measurements taken at separate points gave no reason even to guess that the water temperature varies in a very complex way with the depth. This was established only when continuous measurements were made with the new probes.

It was found that the whole mass of water, from the surface to great depths, is divided into thin layers. The difference in temperature between adjacent horizontal layers may reach several tenths of a degree. These layers are of varying thickness

The Soviet deepwater craft Sever 1, seen here on board a vessel of the Polar Research Centre for Fisheries and Oceanography.

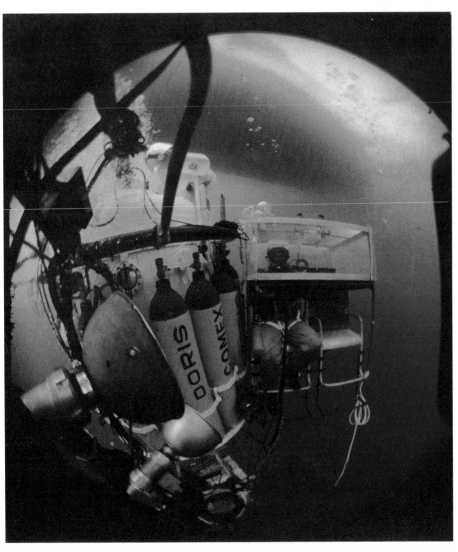

The French firms Doris and Comex have developed original underwater equipment for servicing off-shore oil and gas wells, and for repairing pipelines and communication cables.

and they sometimes exist for several hours and sometimes disappear in a few minutes.

Data of the first measurements of the thin-layer structure of the water strata, made by American oceanographers in 1969, were disbelieved by many scientists. It cannot be, they objected, that the powerful ocean waves and currents do not mix the water until it is completely uniform. But when probing of the water strata at many points in the ocean showed that its thin-layer structure is observed consistently, all doubt was removed. So far, no satisfactory explanation of the phenomenon has been proposed, but it is clear that it is of vital practical importance for underwater acoustic

communication, and also influences the distribution of the mass of plankton in water.

The ocean floor

Only 30 years ago, the use of sounding devices to measure the depth of the oceans was an exceptionally difficult task. It was necessary to lower the heavy sounding lead to the bottom with a steel cable. Errors were common, and soundings were made, as a rule, at points hundreds of kilometres away from one another. This led to the prevailing concept that vast stretches of the ocean floor were flat plains.

A new method of depth measurement, based on the deflection of a sound signal from the sea bed was

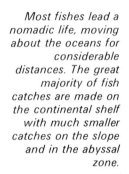

Most fishes lead a nomadic life, moving about the oceans for considerable distances. The great majority of fish catches are made on the continental shelf with much smaller catches on the slope and in the abyssal zone.

first tried out in 1937. The deflection of sound is called an echo, and the new instrument was named an echo sounder. The principle of measurement with an echo sounder is very simple. A special vibrator, attached to the lower part of a vessel's hull, emits pulsating acoustic signals. The signals are bounced back off the ocean floor and are detected by the receiving device of the sounder. The time required for a signal to travel to the bottom and back depends on the depth, and a continuous profile is plotted on a moving chart as the ship sails in a straight line. A series of such profiles, divided by relatively small distances, enable isobaths (contour lines of equal depth) to be plotted on a map which will then show the topography of the ocean floor. Sounding with such sonic depth finders has completely changed previous conceptions of the relief of the sea bed, and given us a much clearer picture of what the ocean floor really looks like.

Extending from the shore under the oceans is a wide strip called the continental shelf. Its depth does not usually exceed 300 metres. A continuous and vigorous transformation of the relief occurs in the upper zone of the continental shelf. The shore retreats from the impact of the waves and, at the same time, considerable amounts of sedimentary matter are formed and deposited. The endless constructive and destructive processes of the waves on the shore lead to the formation of tongues of land and sand bars. These processes lead to lagoonal areas and the partitioning off of the mouths of rivers. In a word, the upper zone of the shelf "lives" a violent geologic life.

The constructive and destructive geological processes seem to become infrequent at depths of 100 or 200 metres. The relief is levelled out and many outcrops of the bedrock can be seen. Demolition of the rock proceeds very slowly.

At the outer edge of the shelf, facing the sea, the surface of the floor begins to slope more steeply, sometimes reaching 40° or 50°. This is the continental slope. Its surface is cut by submarine canyons. These are the sites of intensive, sometimes disastrous processes. Mud and ooze collect on the slopes of these canyons. From time to time the accumulation becomes unstable and a powerful mudstream rushes down at a furious speed to the bed of the canyon, sweeping away sand and stones lying on the slope. This may be a formidable occurrence as one

On the following four pages are seen maps of the ocean floor, showing the major ridges and faults in the earth's crust beneath the surface of the sea.

The rocky surface of the sea bed is ornamented by fan-like coral branches.

Geological processes continuously change the surface of the ocean floor. Lava from underwater eruptions solidifies and forms basalt pieces. These photographs were taken in the Azores area 3,000 metres down.

NORTH AMERICA

REYKJANES RIDGE

OCEAN

MID-ATLANTIC RIDGE

A F R

CAYMAN TRENCH PUERTO RICO TRENCH

SOUTH AMERICA

PERU-CHILE TRENCH

ATLANTIC

ROMANCHE TRENCH

MID-ATLANTIC RIDGE

MID-ATLANTIC RIDGE

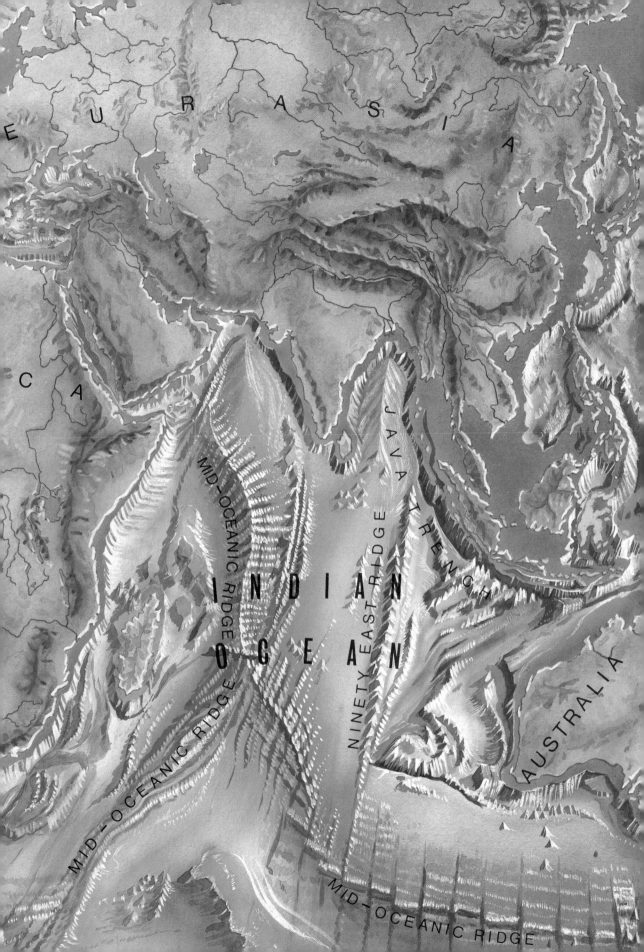

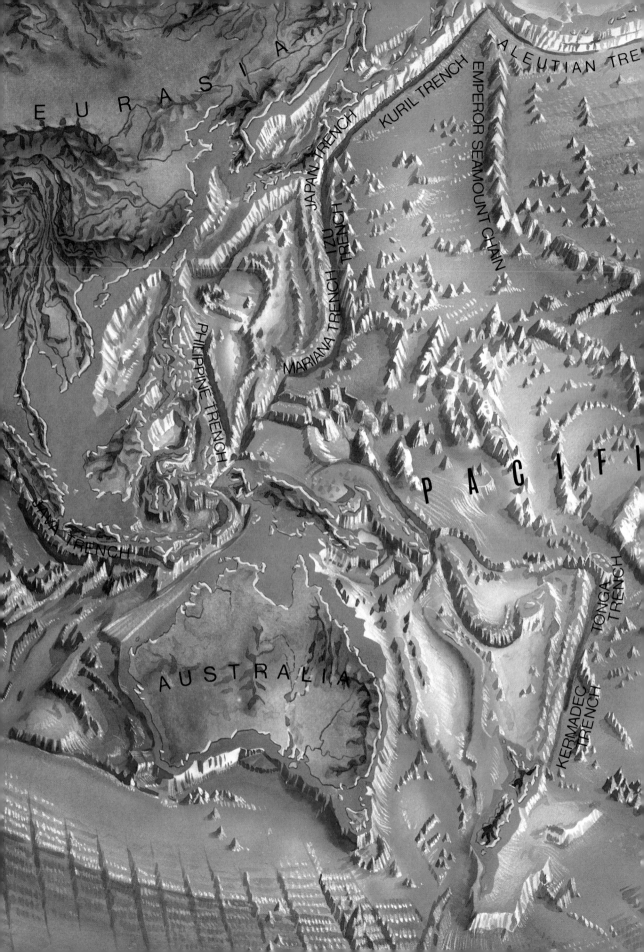

N O R T H

A M E R I C A

MIDDLE AMERICA TRENCH

O C E A N

EAST PACIFIC RIDGE ALBATROSS CORDILLERA

PERU-CHILE TRENCH

S O U T H   A M E R I C A

Coming to the surface is the Soviet remote-controlled apparatus called Manta. *Named after the manta ray it is used for geological and biological research at depths of around 1,500 metres.*

The American submarine Deep Quest, *designed for oceanographical research and rescue operations. In the hull there are two linked spherical compartments for the crew.*

The Soviet one-seater underwater craft Atlas I *is used for biological research on the continental shelf.*

can judge from the rupture of exceptionally strong underwater cables laid across canyons. Such ruptures have been observed on the west coast of Africa, the Atlantic coast of America and many other places.

The ocean floor proper occupies 53 per cent of the sea bed. This was the region that was thought, until recent years, to be a plain. Actually, the relief of the ocean floor is quite pronounced; the floor is divided into immense troughs by sections of the bottom that have been "lifted" to considerable heights. The size of the troughs can be estimated by the fact that the northwestern trough of the Pacific occupies a greater area than all of North America.

Predominant over the troughs themselves is a hilly, rolling relief. Sometimes undersea mountains are found, the height of which may reach five or six kilometres. Occasionally, their peaks may rise above the surface of the ocean. In other regions, the relief of the ocean floor is broken by gigantic, gently sloping bars, hundreds of kilometres wide.

In many places, volcanic cones rise from the ocean floor. Sometimes the peak of a volcano emerges from the surface and an island is born. Some islands of this kind are gradually demolished and disappear again under the sea. Several hundred volcanic cones, broken down by waves and submerged to a depth of up to 1,300 metres, were found in the Pacific.

But volcanoes sometimes have a different evolution. Reef-building coral may make its home at the peak of a volcano and as the volcano is slowly submerged, the coral builds up the reef and, in the course of time, a circular island is formed in the ocean. This is an atoll surrounding a lagoon in the middle. The growth of a coral reef may take a very long time. Certain Pacific atolls were bored to find the thickness of the coral limestone. Thicknesses as great as 1,500 metres were measured. This means that the apex of the volcano sank very slowly and that this took about 20,000 years.

In studying the relief of the ocean floor and geological structure of the oceanic crust, scientists arrived at certain new conclusions. The earth's crust on the ocean floor turned out to be much thinner than on the continents, where the thickness of the solid crust—the lithosphere—reaches 50 or 60 kilometres. In the ocean it does not exceed five to seven kilometres.

It was also found that the composition of the lithosphere on land is different from that in the ocean. On land, under cover of a layer of loose rock and soil—products of the disintegration of the surface of the continent—there is a thick granite layer with an underlying stratum of basalt. There is no granite layer in the ocean and the loose deposits lie directly on the basalt.

Of even more importance was

the discovery of a titanic system of mountain ranges or mid-ocean ridges at the bottom of the ocean. The mountain system of mid-ocean ridges stretches through all the oceans for a distance of 80,000 kilometres. In size, the undersea ranges can be compared with the highest mountains on land, such as the Himalayas. The crests of the undersea ranges are riven lengthwise by deep gorges which have been named rift valleys, or simply rifts. Their extensions can be traced on land as well.

Scientists realized that the underwater global system of rifts is a phenomenon of vital significance in the geological development of our planet. A period of careful study of

On board the French bathyscaphe Archimède the crew prepares to submerge.

sea urchin starfish  crab  sea cucumber sea squirt sea anemone

sea urchin

 eolid mollusc

 hermit crab in mollusc shell

 sea lily

 cuttlefish

scallop

sponge

the system of rift zones was begun and the data obtained were so unexpected that they led to drastic changes in the conception of the earth's geological history.

The ruling hypothesis up to the turn of the century was that our planet began as a fiery mass of molten matter. It was assumed that a lengthy process followed in which the matter cooled and condensed. This led to the formation of a solid crust on the earth's surface which, incidentally, did not lose a certain amount of elasticity. Therefore, as the earth continued to cool, the crust contracted (all bodies contract on cooling), and folds were formed that became the existing mountain

The cowling of the French submarine Cyana *is gently lowered into place. One of the most modern underwater research vessels, it can submerge to a depth of 3,000 metres.*

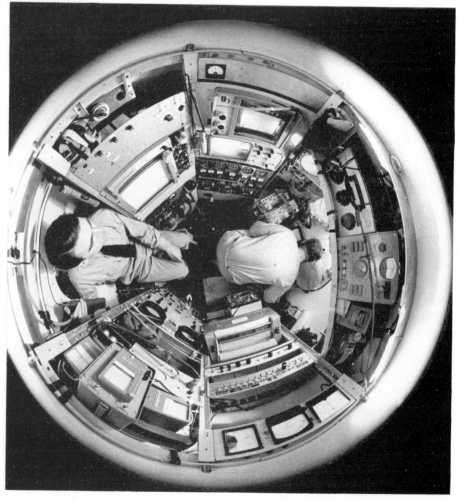

Within the gondola of the bathyscaphe Trieste *there are so many scientific and navigational instruments to attend to that the crew has little time to think of its cramped quarters.*

systems on land. Usually, to make it more convincing, advocates of the hypothesis cited the analogy of a baked apple on whose surface deep seams are formed. This hypothesis, known as the contraction hypothesis, was clear and simple; it possessed exceptional inherent orderliness and seemed to explain all the known facts in a satisfactory way.

But at the end of the 19th century, the English physicist Osmond Fisher published a work called *Physics of the Earth's Crust* in which he contended that the contraction theory cannot explain the existence of considerable horizontal movements of masses of land. Fisher assumed that the earth's evolution proceeds under the action of currents in the molten magma on which the solid crust lies. These currents cause expansion and cracks along the centre of the oceans with an outflow of basalt magma through the fissures. Expansion of the oceans leads to the sinking of the ocean crust under the continents and islands. This is why such frequent and violent earthquakes occur at the periphery of the Pacific.

Fisher's investigations contained the rudiments of all modern concepts of the earth's geological evolution. His work, however, was not noticed at the time, and the crushing blow dealt to the contraction hypothesis is ascribed to the German scientist Alfred Wegener, who formulated his theory of continental drift in 1912.

Wegener succeeded in reconstructing a picture of the creation of the earth and its evolution for the last 200 million years. He contended that the composition of the earth's crust differs essentially on the continents and under the oceans. Predominant in the continental crust are granite and gneiss rock, consisting mainly of silicon and aluminium.

Sea anemones are often found on the ocean floor at depths of up to 5,000 metres.

Scientists still know very little about the life of fishes that live in the ocean depths.

Acorn worms belong to a primitive group of animals called hemichordates. They live on the sea bottom burying themselves in mud and sand.

Some treasures of the ocean bed – a field of ferromanganese concretions. Such ore fields sometimes stretch for hundreds of kilometres.

This crust is lighter than the oceanic crust whose chief components are sial (silicon-aluminium) floating in an underlying viscous base called sima (silicon-magnesium).

According to Wegener's reconstruction, the initial single continent of Pangaea began the evolution. Then a wide split occurred into which the ancient ocean Tethys penetrated, dividing Pangaea into two parts: Laurasia in the north, uniting North America and Eurasia, and Gondwanaland in the south, uniting South America, Africa, India, Australia and Antarctica into a single unit. Some time during the last 100 million years, Laurasia and Gondwanaland were further divided. Each unit drifted apart and the continents finally occupied their present positions. But this brilliant hypothesis had one weak point. Wegener could find no satisfactory reason why the continents had drifted apart in the first place.

During the International Geophysical Year, the attention of scientists was again drawn to the

The American research vessel Glomar Challenger. *Instead of the conventional ship's masts and superstructure there is a rig on board for drilling operations at sea.*

problem of the geological evolution of the earth. They revived the half-forgotten continental drift hypothesis proposed by Wegener.

The contours of the continents separated by the Atlantic Ocean were carefully compared. Geophysicist Sir Edward Crisp Bullard superimposed the contours of Europe and North America, and Africa and South America, not along the shore lines, but along the middle of the

continental slope, approximately at the 1,000-metre isobathic line. Along this line the outlines of the two shores coincided so closely that even those who had previously disagreed could not doubt the horizontal displacement of the continents.

Especially convincing were data obtained during geomagnetic surveys in the area of the mid-ocean ranges. It was found that the outflowing basalt lava is evenly dis-

placed on both sides of the ridge on a range. This was direct evidence of the spreading of the oceans as the earth's crust separates in the vicinity of a rift valley and, consequently, of continental drift.

Deep drilling in the ocean floor, performed from aboard the United States ship *Glomar Challenger*, has again confirmed the spreading of the oceans. In the first place, no rock could be found with a geological age

From such ships as the Glomar Challenger *scientists bore deep wells in their efforts to unravel the secrets of the earth's origin and structure and determine the distribution of minerals.*

located. Calculations showed that 96 per cent of the energy of all earthquakes was released in the regions where there is an underthrust of the ocean crust into the continents.

The basis of the new theory of plate tectonics is the concept of the solid crust of the earth, the lithosphere, being divided into separate plates. These plates are subject to horizontal movement. The forces that move the lithospheric plates are developed by the convective currents, i.e. currents deep in the fiery molten matter of the earth. Drifting of the plates away from each other is accompanied by the formation of mid-ocean ridges with yawning fissures of rift valleys at their crests where basalt lava flows out into the ocean. In other regions the plates approach each other, collide and interact. In these collisions, as a rule, one plate moves under the other. Such modern underthrust zones are known at the periphery of the oceans where troughs and volcanoes are formed and earthquakes are frequent.

The conceptions held today of the internal structure of the earth and of the processes that occur in its interior are based, for the most part, on the cosmogonic hypothesis of the Soviet Academician O. Yu. Schmidt. He contended that the earth, like the other planets of the solar system, was formed by the cohesion of the cold matter of a dust cloud. Further growth of the earth took place as it captured meteoric matter in passing through the dust cloud that once surrounded the sun. As the planet grew bigger and bigger, the heavier (iron) meteorites sank and the lighter ones, of rock, floated to the top. The separation or differentiation of these meteorites was such a powerful process that inside the planet the matter melted and divided

With increasing confidence man is exploring the vast expanse of the oceans. Prospecting for off-shore oil and natural gas in many parts of the world has led to such typical scenes as this "forest" of oil derricks rising out of the sea off the coast of the United States.

exceeding 150 million years. This means that geologically the oceans are very young. Secondly, the absolute geological age of the basalts increases consecutively in the direction from a rift valley to the periphery of the ocean. In other words, the farther basalts are from the place of their origin, the more ancient they are.

An explanation was also found for the higher earthquake and volcanic activity at the periphery of the oceans. Here the ocean crust of the earth protrudes under the edge of a continent and deep troughs are formed, such as the well-known trenches encircling the Pacific, and volcanic islands. Here is where the origins of deep earthquakes are

into a heavier part, difficult to melt, and a lighter, easily melted part. Simultaneous with these processes was the radioactive heating of the earth's interior portions. Naturally, such intensive interaction led to the formation of the heavy inner, or central, core, the outer core, and the inner and outer mantles. Geophysical data and calculations indicate that concealed in the depths of the earth is boundless energy capable of vigorously transforming the solid crust we call the lithosphere.

There actually is a continuous differentiation of matter in the outer mantle, and its more easily melted part penetrates to the surface of the lithosphere in the form of basalt lava. Thus, the theory of lithospheric plate tectonics does not conflict with the previously accepted theories of the earth's evolution. Concepts of the structure of the earth's crust and its dynamics are therefore gradually becoming clearer.

Life in the oceans

The field of marine biology—life in the sea—has probably had the most discoveries of any field of study of the oceans. Many hundreds of animals, previously entirely unknown, have been found in the world's oceans in the last 25 years.

But a new general conception of the development of life in the ocean resulted in other discoveries as well. Life was first found at great depths in 1949. This discovery was made

Here is a rich haul of fish and Kamchatka crabs caught in the Sea of Japan. Thousands of vessels ply the seas in search of fish and conservation of food-fish resources is vital.

almost simultaneously by an expedition on the Soviet research ship *Vityaz* and a Danish expedition on the ship *Galatea*. Before this discovery the question of the existence of life at the enormous pressures which exist in the depths of the sea was a subject for heated discussion.

Opportunities soon arose for a detailed investigation of the anatomic and physiological features of

The speed at which the sailfish can swim is due partly to its streamlined sword or bill which helps to reduce water resistance. The same design principle is employed in the nose of supersonic aircraft.

Colony of tropical fluorescent coral.

deep-sea organisms and, what is of prime importance, of determining the zone of maximum depth or, as scientists say, the ultra-abyss to which living matter of the ocean extends.

Years of investigations into the biology of the oceans enabled voluminous data to be collected and much turned out to be new and unexpected. It was found, for example, that zoo plankton in search of planktonic food, make regular vertical migrations. A study of the distribution of the phytoplankton showed that only the periphery of the ocean and zones where deep water rises were productive. In general the ocean is nothing but a lifeless desert in which only coral islands, or atolls, appear to be rare and very small oases of abundant life.

Paracanthus hepatus, the surgeonfish, is one of the colourful creatures found in the Indo-Pacific coral reefs.

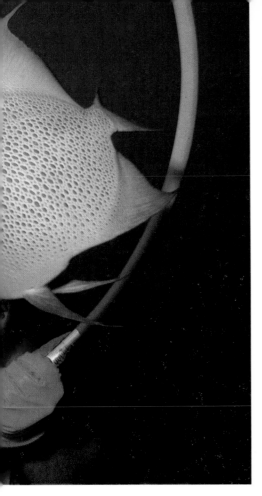

Centre left: This angelfish shows little fear when taking food from an aquanaut's hand, demonstrating that to some extent fishes can be tamed.

valuable resource, containing dozens of chemical elements, is the sea water itself. About one third of all the common salt used in food is obtained from the ocean. Sea water also yields bromine and iodine, as well as hundreds of thousands of tonnes of magnesium.

Today, underwater deposits of oil and gas, placer deposits of titanium and gold, iron-manganese concretions (nodules lying on the ocean floor) and tin, coal and diamond deposits, are all being rigorously mined and utilized. In his development of oil and gas wells, man has already gone beyond the limits of the continental shelf. The world's oceans are also gigantic accumulators of inexhaustible stores of energy. The first electric power stations based on tidal energy have already been built.

A tree-like colony of coral polyps belonging to the Class Anthozoa, a name which comes from the Greek and means flower animals. These coelenterates are aptly named as they look like plants rather than animals and spend their lives fixed to the sea bed.

In appraising the knowledge that man has acquired, it can be said that we have taken only the first steps in our understanding of the oceans. Much has still to be found out about the nature of the sea, and science has yet to travel the long and difficult road of research into the basic laws governing its dynamics and development.

Exploitation of the ocean's resources should always be undertaken with care. Each practical activity, be it fishing or the extracting of oil or minerals from the sea bottom, should be based on a precise knowledge of the laws of nature that pertain to the sea, with an awareness of the eventual as well as the immediate consequences.

Many of the ocean's mysteries are still to be unravelled and many discoveries have yet to be made by scientists. The following chapters tell about those courageous people who do their work in the sea—a medium hostile to man—and stand face to face with the oceans.

A significant discovery made recently was that up to 30 per cent of the zooplankton by weight in the food ration of certain crustaceans turned out to be clusters of bacteria. This means that the first link in the food chain is made up of bacteria as well as phytoplankton.

Biological research in the oceans acquired special significance in connection with the danger of destroying food-fish resources. The world catch of fish and marine animals has reached 70 million tonnes per year. If this figure passes the 100 million-tonne mark, experts contend that irreparable damage will have been done.

Mineral resources of the oceans

The oceans are also rich in chemical and mineral resources. One most

CHAPTER 2

DIVING
TO THE
SEA FLOOR

Man employs various methods in his quest for the sea's treasures. Hundreds of ships, some of them actual floating institutes, are engaged in revealing the mysteries of the ocean depths. Huge platforms, either fixed or mobile, are used, for drilling boreholes in the ocean floor, when prospecting for oil or for geological exploration. The surface of the ocean, however, is hardly the best place from which to conduct investigations. The exact information needed by scientists can be obtained in many cases only by a descent into the depths of the sea — by walking along its floor.

From Aérophore *to aqualung*

In Jules Verne's novel *Twenty Thousand Leagues under the Sea*, Captain Nemo, his sailors and Professor Aronnax walk along the sea bed carrying on their backs an apparatus which contains a sufficient supply of breathable air. As Captain Nemo explains to the professor, the design of the equipment was simply based on the Rouquayrol-Denayrouze apparatus, invented by two Frenchmen in the 1860s.

Many readers may think that the underwater breathing apparatus used by Captain Nemo and his men was only the brilliant fantasy of a famous writer. They may also think that Verne gave imaginary names to

the inventors of the remarkable diving equipment. But this is not so.

In 1825, William James, an Englishman, proposed a new self-contained device for diving. His invention consisted of a metal reservoir filled with compressed air, the used gas being exhausted from the diving helmet through a special valve. Forty years later, two Frenchmen produced a new invention. Mining engineer Benoît Rouquayrol and naval officer Auguste Denayrouze modestly declared that they had tried only to improve the James apparatus. But, in fact, they had devised an entirely original and wonderful apparatus.

The *Aérophore*, as the new apparatus for breathing under water was called, consisted of a metal cylinder filled with compressed air and strapped to the diver's back like a knapsack (in contrast to the reservoir of the English inventor which was worn on the belt), and a box-like mask with a glass porthole that covered the eyes, nose and mouth. But the main feature of the Rouquayrol-Denayrouze *Aérophore* was its regulating valve, an ingenious device consisting of a membrane with one side subject to the pressure of the water and the other to that of the compressed air being delivered to the diver for breathing. This automatically regulated the pressure of the air being inhaled to the

The French novelist Jules Verne anticipated many scientific possibilities in his stories. The setting of one of them, Twenty Thousand Leagues Under the Sea, *is of particular interest because of the underwater breathing devices it describes. In this 19th-century engraving Captain Nemo and his companions walk on the ocean floor surrounded by giant seaweeds and jellyfish.*

pressure of the water surrounding the diver.

Aérophores, however, never came into wide use. There were many reasons for this, the chief being, perhaps, the lack at that time of reliable and sufficiently capacious cylinders and high-pressure compressors to fill them with compressed air. As we know, the creator of the *Nautilus* solved these problems in his novels, but the situation was somewhat different in real life.

Intended at first as a self-contained apparatus, not connected by a hose to a ship, the Rouquayrol-Denayrouze device was in practice only partly self-contained because of the extremely limited capacity of its reservoir. Periodically, air had to be forced down to the reservoir through hoses (the feature that Captain Nemo did not like). Besides, no convenient masks or swim fins then existed.

Several decades later the *Aérophore* was succeeded by an apparatus developed by Yves Le Prieur. But the membrane of its valve proved to be insufficiently sensitive to changes in the surrounding pressure. Consequently, the diver was obliged to regulate the flow of fresh air by hand. Le Prieur's apparatus was improved by Georges Comheines. He had two cylinders of compressed air instead of one. But still man did not feel himself as free as a fish in water. This dream was only made possible by Jacques-Yves Cousteau.

In 1943 Cousteau met Emile Gagnan who knew nothing about diving but was an expert on industrial gas equipment. Gagnan had proposed that gas generated from coal be used as fuel for automobile engines since petrol was hard to get during the war years. Demonstrating his device for feeding gas to the engine, he once told Cousteau, "The

problem is somewhat the same as yours."

And so the Cousteau-Gagnan regulating valve was developed. It is the most essential part of the aqualung, which has enabled man to conquer the seas. The principle of the valve is very simple. It is a two-stage demand regulator which, in the first stage, reduces the pressure of the air flowing in from the cylinders of compressed air. In the second stage, the valve automatically regulates the pressure of the air to suit the depth of the dive (it was not until the 1950s that a single-stage valve was developed).

Thus the invention of the aqualung, more properly called a scuba (*s*elf-*c*ontained *u*nderwater *b*reathing *a*pparatus), enabled the hose and lifeline, which had confined the helmet diver for over a century, to be severed. The skin diver thus gained the freedom and mobility of an amphibian.

The "Underwater Age"

The advent of the aqualung, or scuba, marked the beginning of a new era in the study of the sea depths. But, as has frequently been the case in history, the new apparatus did not immediately gain recognition.

The march of events was accelerated by the initiative and energy of Cousteau and the friends who shared his ideas. Cousteau bought an old navy mine-sweeper, written off at the end of World War II. After re-equipment, this military vessel became a floating laboratory, the world-famed research ship *Calypso*, which faithfully serves Cousteau to this day, sailing the seas and oceans of the earth.

With more and more assurance the aqualung is conquering the world of the deep and so far nothing

Sea anemones are most common in shallow inshore waters where they attach themselves to rocks, although some species live in deep waters. Their muscular cylindrical bodies end in an oral aperture surrounded by tentacles. Frequently they attach themselves to the hulls of ships and are taken on long journeys.

can take its place. Simple, inexpensive and convenient, it has become the indispensable tool of those who investigate the oceans. It has also made it possible for hundreds of thousands of underwater tourists in all parts of the world to enjoy the wonders of the sea.

"Menfish" at work

In the years that have passed since Cousteau and Gagnan met, researchers of many countries have conducted thousands of expeditions into the depths of the world's oceans. A great many secrets of the underwater world have been disclosed with the aid of aqualungs.

Some of the richest oil and gas fields and tin deposits have been explored, and their extraction from the sea bed began many years ago. Scientists equipped with aqualungs and swim fins, diving fearlessly into the depths of the sea, have made valuable contributions to science.

In the Arctic huge, drifting ice fields extend for hundreds of kilometres. Small wonder that for centuries these severe regions were inaccessible to seafarers and explorers. In 1937, Ivan Papanin and three colleagues first made camp on an ice floe near the North Pole. The drifting ice field on which they settled simply roamed the Arctic Ocean, zigzagging its way at the will

Jacques-Yves Cousteau, one of the pioneers of underwater exploration and the inventor, with Emile Gagnan, of the aqualung.

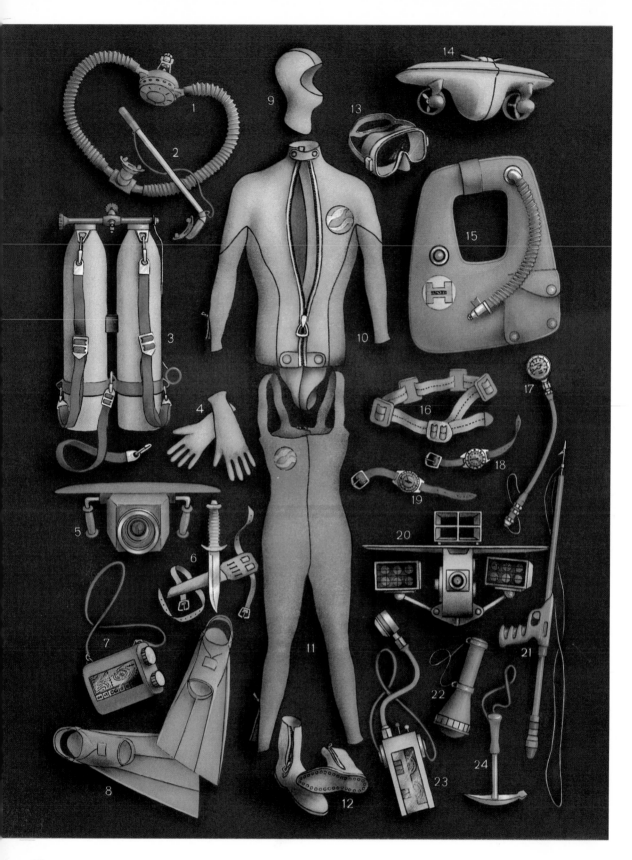

of the currents it encountered. Even under such unusual and harsh conditions, Papanin's team of scientists conducted extensive oceanographic observations in the polar waters.

Two decades later, in August 1958, the American atomic submarine *Nautilus* made an underwater traverse of the ice and reached the North Pole. Soon Soviet submarines appeared in these high latitudes. Breaking the polar ice pack, they surfaced at the North Pole.

Next came the turn of the skin-diving explorers. V. Savin, Yu. Pyrkin, V. Grishchenko and their colleagues, scientific workers of the Arctic and Antarctic institutes of the USSR Academy of Sciences, descended in rubber suits into the depths of the Arctic Ocean, at the spot where the imaginary axis of rotation of our planets intersects the surface of the globe. Based at the Soviet drifting research station North Pole 18, these divers made over 400 dives with aqualungs through cracks and holes in the ice to depths of up to 50 metres.

Meanwhile, investigators were also busy in Antarctic waters. In October 1965, the Soviet diesel-electric ship *Ob* left its moorings in the port of Leningrad for its next regular voyage to the shores of distant Antarctica. A group of skin-diving biologists were on board, together with other scientists who were also members of the polar expedition.

When one skin-diver emptied his trophies out of his collecting bag, it was an unusual sight. Purple sea urchins waved their needles; gigantic water worms, almost a metre long, wriggled convulsively. Side by side lay bright red and violet starfish, many-hued sea anemones and bunches of soft coral. These exotic inhabitants of the Antarctic sea

seemed to be strangers from another world, seen against the background of the endless, monotonous land of ice and snow. The discovery was so unexpected: a rich diversity of life hidden under the age-old covering of ice.

In the decades spent by the famous French explorer Jacques-Yves Cousteau in his underwater expeditions, his cameramen made many interesting motion pictures about the world of the ocean depths. They also filmed in freshwater

basins all over the world, notably in Africa. Ingenious methods were invented by the French skin divers to enable them to film hippopotamuses frolicking under the water in Lake Tanganyika. Despite their impressive size and seeming clumsiness, these animals are first-class underwater swimmers. A rubber hippopotamus was made

Opposite: Some of the underwater swimmer's standard equipment.
1. breathing apparatus;
2. breathing tube;
3. aqualung;
4. rubber gloves;
5. waterproof camera;
6. diver's knife and sheath;
7. roll map of sea bed section;
8. flippers;
9. helmet of thick sponge rubber;
10 and 11. swimmer's neoprene suit;
12. rubber boots;
13. face mask and goggles;
14. powered apparatus to assist underwater swimming;
15. emergency life jacket (a press on the button immediately inflates the jacket and propels the diver to the surface);
16. weighted belt to neutralise buoyancy;
17. manometer (to indicate the gas pressure in an aqualung);
18. depth gauge and watch;
19. compass and thermometer;
20. cine camera;
21. underwater gun;
22. flashlight;
23. tape recorder and microphone;
24. hammer.

Left above: Gorgonocephalus is a basket star related to starfish.

Left below: Hermit crab makes its home out of the hollow shell of a gastropod mollusc.

Opposite: As a safety precaution, divers work as a team when exploring an underwater grotto such as this.

There are about 5,500 living species of echinoderms. They have a great variety of form and colour and divide broadly into two groups, those which live mostly in fixed locations and those which are free to move about. Among the echinoderms are the sea urchin, one of which Heterocentrotus mammillatus *is seen here. It has long triangular rod-like spines.*

with the motion picture camera arranged in its head, the nostrils serving as portholes. Hidden in the body of the pneumatic hippo, Philippe, Jacques-Yves' son, stealthily approached the herd and shot the film without difficulty.

Other skin divers from Moscow, including physicist Victor Suetin, had an unprecedented adventure. They descended with aqualungs into the crater of a volcano under water. "It was a dive I shall never forget," related Suetin. "The volcanoes of the Kurile range, topped with pointed cones, stretch out in a long chain. Our group of skindiving investigators climbed one of the dormant volcanoes to study the water of the lake filling the crater. The air was heavy with

sulphurous fumes, hot gases spurted from fissures in the rocks, and the water boiling near the shore was covered by a black volcanic foam. Descending into this natural witches' cauldron, we took samples of the soil at the bed of the volcanic lake, sampled the water and the gases, and measured the temperature at various depths."

Astronauts also find use for the aqualung. Their training includes dives in water in order to experience hydraulic weightlessness.

Nitrogen narcosis

"My three colleagues and I dived to a depth of 63 metres. Then suddenly I became confused. I soon felt that I could no longer continue the descent and, with great difficulty, began to pull myself upward along a line hanging from the surface. My condition became worse and worse. At times it seemed that somebody was stealing up behind me and was about to tear off my mask. Then there was more light, the surface was near and I suddenly wanted to swim back down into the depths and remain there for ever."

In the experiment recalled here by David Robinson, an Australian diver, the purpose was to descend to a depth of 75 metres using an aqualung charged with ordinary compressed air. What had happened to the divers? They had become victims of deep-water intoxication, termed *L'ivresse des grandes profondeurs* (Rapture of the great depths) by Cousteau. It is due to nitrogen which, as is known, makes up 78 per cent of atmospheric air. At the high pressures found at depths of over several dozen metres, this gas becomes a dangerous narcotic.

Deep-water intoxication, or nitrogen narcosis, has cost the lives of hundreds of helmet scuba divers. One of Jacques-Yves Cousteau's experiences of this danger almost ended fatally. Practice has shown that an aqualung, charged with ordinary compressed air, as well as a diving suit to which air is pumped from the surface, can only be used regularly at depths up to 40 or 50 metres, the maximum dives being 70 or 80 metres, depending upon the individual characteristics of the diver.

Nitrogen narcosis became 'a formidable obstacle in reaching greater

depths. To avoid it, various artificial breathing mixtures were proposed. The best of these turned out to be a mixture in which nitrogen was supplanted by the inert gas helium. The first experimental helium-oxygen breathing mixtures were a success and enabled divers to descend to depths substantially greater than those reached by a diver breathing ordinary compressed air.

But treacherous deep-water intoxication is not the only enemy of men diving to considerable depths. In breathing under high pressure the body absorbs nitrogen or helium. The greater the depth of the dive and the longer the diver stays under water, the more his blood, muscles and tissue are saturated with gas.

The blood of a diver is sometimes likened to champagne. Wine, saturated by a gas and corked under pressure in a thick-walled, heavy bottle, behaves very peacefully. But as soon as the bottle is opened, the champagne—due to the sudden drop in pressure—instantly begins to foam with thousands of gas bubbles.

This, roughly, is the same process that occurs when a diver comes up to the surface too rapidly. The only difference is that the gas goes into solution in the blood instead of being released into the air. Then the bubbles of gas clog the blood vessels, causing caisson disease, more commonly known as "the bends", the scourge of all divers.

The clogging of the finer blood vessels impairs blood circulation, leading to a lack of oxygen to the brain. Air bubbles in the spine or brain paralyze the appendages, and if the nitrogen bubbles reach the heart they cause instant death. It has been established, however, that caisson disease can be avoided if the diver comes up slowly, gradually reducing the pressure of the gas mixture. Then the gas is exhausted into the lungs, instead of the blood, and is discharged through the valve of the diving suit or aqualung into the sea. Helmet and scuba divers call this gradual reduction in pressure, decompression.

Decompression was simplified somewhat when the undersea lift, or submerged decompression chamber, was developed. It is actually a diving bell, has a hatch underneath and is suspended from a windlass or jib crane on board a surface tender. When the diver finishes work, he enters the lift and seals the hatch. While the lift is slowly raised, the diver goes through the decompression process, spending his time in a dry space.

Divers encounter other difficulties when they are submerged to great depths. For example, the high density of the gas leads to over-exertion of the muscles employed in breathing. This, in turn, impairs air

A colony of mussels. They belong to a group of molluscs which includes oysters and scallops. The meat of the mussel is a popular food in many countries and many mussel-breeding "farms" have been set up to cater for the demand.

circulation in the lungs. The helium-oxygen mixture or, simply, heliox, renders an important service here too. Helium has a lower density than nitrogen and comes out of solution in the body considerably quicker and is exhaled.

Many breathing mixtures, containing various gases in different proportions, have been tested in the sea and in pressure chambers on land. Such chambers can simulate submersion to great depths. As a result, a breathing mixture can be prescribed for any depth and for any length of time at this depth.

Diving records

Owing to the improved mixtures that have been devised, man has achieved some impressive dives in his prolonged duel with the sea depths. In 1937, Max Nohl, an American, used a diving suit he had designed himself to reach the record depth of 135 metres. Two years later, Soviet divers, Leonid Kobzar and Pavel Vygulyarny, broke this record. They descended 157 metres and were the first to return alive from such a depth, unheard of at that time. Both the American and the Soviet divers used the heliox breathing mixture.

Several years later, in 1946, another American, Jack Browne, set a new record. Not in the open sea, but in a hydraulic pressure tank on the shore in which a dive of 168 metres was simulated. Two years after this, an Englishman, William Bollard, breathing heliox, dived to almost the same depth under natural conditions. His dive was only about three metres less than Browne's, whose depth record was eventually surpassed.

It is estimated that there are 1,600 living species of starfishes. They are echinoderms, related to sea urchins. One useful attribute they share with certain crustaceans is the ability to regenerate an arm should it be damaged or broken off.

CHAPTER 3

LIVING
ON THE
SEA BED

Dreams of undersea settlements could be realized when the so-called saturation effect was discovered. It had already been mentioned that a man under water absorbs an inert gas, nitrogen or helium. It was established that a moment comes when the tissues of the body do not absorb any new doses of gas. After that moment, the time required for decompression in ascending will be the same no matter how much longer the man spends under water—an additional two minutes, an hour, a day or even a whole month. The length of decompression, in this case, depends only on the depth of the dive and how well the breathing mixture has been chosen and prepared.

To test this discovery, investigators conducted hundreds of experiments with animals. Dives to various depths were simulated in pressure tanks. The physiological and psychological effects of high pressures on man were studied, and experiments were conducted with volunteer divers. Finally, the time came to test the idea under natural conditions.

The first to achieve success was the American engineer Edwin Link. His underwater quarters consisted of an airtight vehicle or gondola, with the hatch underneath. As the gondola was submerged, the pressure of the air inside was kept equal

to that of the water outside. When the hatch was opened, the compressed air kept out the water. The inhabitant of this underwater home stepped out or, to be exact, swam out over its threshold with no limits on the time spent in the open sea. Returning home, he could rest, eat and then go out of his premises again.

In August 1962 Edwin Link personally tested his underwater dwelling off the shores of France, near Toulon. He first made several short dives to depths up to 18 metres. A few days later one of the most experienced of Belgian skin divers, Robert Stenuit, came to help Link. Quickly becoming familiar with the apparatus, Stenuit repeated the test dives at a small depth. On September 6 Link gave the command to start the key experiment and the aluminium capsule with Stenuit aboard was lowered to a depth of 60 metres.

Despite some overcrowding, living conditions were not at all bad in the underwater house: there was light and warmth, which are the primary needs. Electric power for heating and lighting came through a cable from the attendant surface ship. From there compressed "artificial" air, seven times as dense as ordinary air, was delivered in a continuous stream through hoses. The air contained 94.6 per cent

Jacques-Yves Cousteau aboard the research vessel Calypso. Cousteau has often described the ship as his true home and for over 30 years the names Cousteau and Calypso have been inseparable.

helium and only 3.6 per cent oxygen. The dweller could talk by telephone with the shore observers, sharing his impressions. But at that great depth, in air thickened by helium, speech became inarticulate and unpleasant to the ear, owing to the effect of gases on the vocal chords. So the telephone was abandoned and all messages were sent in Morse code by a telegraph apparatus.

The first day of life in the depths of the sea came to an end. Stenuit left his dwelling several times to become acquainted with the surrounding world. At night he again made brief trips into the ocean. It was intended that he should spend one more day in the capsule and there seemed no difficulty in this plan. Then suddenly a heavy leakage of helium was detected. Link ordered his men to start raising the underwater dwelling. Obeying instructions, Stenuit securely closed the hatch.

The capsule with its inhabitant

was raised to a depth of 30 metres and then, according to a schedule calculated beforehand, the pressure inside was slowly reduced. By the evening of September 7 the underwater house was aboard the ship. Decompression started on the fourth day. Early in the morning of September 10, Stenuit opened the hatch and came out of the dwelling in which he had spent 92 hours, 30 minutes. So ended the first experiment with an undersea dwelling, something which had once seemed a fantasy.

Jacques-Yves Cousteau

Only a few days passed before a new underwater dwelling appeared at the bottom of the Bay of Marseilles in the Mediterranean Sea. Its founder was Jacques-Yves Cousteau and its tenants were the associates of the famous explorer, the expert divers Albert Falco and Claude Wesly. At noon on September 14, 1962, Falco and Wesly climbed down the ship's ladder of the *Calypso* and disappeared under the water.

In spite of the unusual experiment being conducted, the underwater dwelling seemed quite commonplace and looked like a railway tank car turned upside down. Having a considerable reserve of buoyancy, the house was held in place by anchors at a depth of ten metres. At the surface, from the attendant ships *Calypso* and *Espadon*, hung flexible pipes, hoses and cables. They delivered fresh air at a pressure of two atmospheres, electric power, and hot and cold water.

The furnishings of *Diogenes*, as the divers named their house were quite simple. They were supplemented by instruments for measuring humidity, temperature and pressure, a system for recording the water level in the entrance shaft, and a closed-circuit television camera for observing the aquanauts.

On the first day, Cousteau visited the underwater dwellers and made sure that they were happy with their situation. In fact, the aquanauts were in an elated mood: with one step they could swim leisurely into the greenish sea water around them for as long as they liked. The pressure in *Diogenes* was always equal to the external pressure. Consequently, water from outside could not penetrate inside the dwelling and the entrance hatch, which the aquanauts called their "liquid door", was usually kept open. The only thing that displeased the aquanauts was the doctor, who came to visit them in their underwater dwelling, making thorough examinations lasting up to two and a half hours and distracting them from their work.

On the third day, however, the attitude of *Diogenes'* inhabitants changed abruptly. Both became silent. They listened to news and directions from the surface with indifference, asking no questions. They had no heart to do anything and they could not sleep at night.

On the fourth day, Cousteau again visited Falco and Wesly. He told them that the evening visit by the doctor had been cancelled. The news enlivened the men. A sudden change seemed to take place in their spirits. They both became noticeably more cheerful. Their period of acclimatization in the undersea world had come to an end and they had finally settled down.

As the days passed, life in the sea depths captivated Wesly and Falco more and more. In investigating the life of fishes, the aquanauts built them labyrinth-like sanctuaries of blocks. To their great satisfaction, Falco and Wesly found that their labours had not been in vain. The fish moved readily into the town constructed for them. The inhabitants of *Diogenes* called it their underwater ranch.

The expedition in the Bay of Marseilles, named *Précontinent 1*, ended on the seventh day. On September 21, 1962, in the daytime,

The thresher shark is also known as the sea fox. It feeds on smaller fish such as herring and will sometimes move into a shoal and thresh the water with its tail fin in order to stun its prey.

Wesly and Falco made their final circuit around the roof of *Diogenes* and swam unhurriedly to the surface. As with the experiment by Link and Stenuit, the expedition *Précontinent 1* was brilliant and incontestable proof of the possibility of an active and useful life for man in the depths of the sea.

The thornback ray lives on the muddy bottoms of inshore waters.

Aquanauts at the Roman Reef

A new roomy dwelling for aquanauts was now under the keel of the *Calypso*. This time the house was located at the bottom of the coral lagoon Shab Rumi, the Roman Reef, in the Red Sea. The name of this comfortable dwelling made of duralumin was *Starfish*, which it resembled in shape. *Starfish* was anchored at a depth of ten metres and was occupied simultaneously by seven French aquanauts. Later two of them moved to another underwater dwelling called *Little*

House, which was held at a depth of 26 metres.

Yet another unique building was located in the sea. It was called *Sea Urchin* and served as a hangar for the submarine *Denise*. Returning from regular trips to the depths, the submarine would slip between the steel struts of *Sea Urchin* and stop. An electric crane carefully gripped the submarine and lifted it into the hangar. Here, as in the other underwater houses, it was dry and the pilots of the *Denise* came out into the same surroundings as the aquanauts.

The underwater settlement in the Red Sea was a well-planned complex of living and working premises. Every effort was made to create normal living conditions for the men, the more so because they were to stay for quite a long time. Jacques-Yves Cousteau said, "Our aim was to achieve effective and vital activity under water at a depth

Old friends and associates discuss their next moves aboard the Calypso. *(From left to right) André Laban, Frédéric Dumas and Jacques-Yves Cousteau.*

of at least 200 metres for a period of a whole month. If we succeeded we should present mankind with the means of mastering the reaches of the undersea continental plateau." The experiment was named *Précontinent 3*.

Stupendous efforts by all the participants of the expedition were required to erect the houses on the sea floor. The heat was blistering and it was difficult to breathe. Instead of a long-awaited breeze, a searing wind blew from the shore carrying clouds of scorching sand. In this baking heat, the men unloaded all the necessary structures and the lead ingots used for ballast. It was no easier for the men working under water. To prepare the construction site, it was necessary to root out the coral and to level off the area. Moreover, the anchors holding the depot ship *Rosaldo*, which was to supply the village under the sea, had to be accurately located. Finally, the divers had to

set up the underwater houses *Starfish*, *Little House* and *Sea Urchin* which, being hollow and metallic, had great buoyancy, and the hangar for the submarine *Denise*.

On June 15, 1963 the ceremonial settlement of *Starfish* took place. From the very beginning, the undersea dwellers and divers carrying messages to them tried hard not to disturb the peace and natural life rhythm of the coral reef. Fish for food were speared several kilometres from the underwater settlement. The aquanauts tried to get on friendly terms with the reef dwellers; they fed the sea perch and trigger-fish, and even such dangerous predators as moray eels and barracuda. Their particular friend was Julie, a barracuda which displayed rare friendliness in its contact with people.

Dwellers of the Little House

On July 5, three weeks after the housewarming party at the floor of

A crew member of Starfish *swims across the path of the submarine* Denise *in the coral bushes off the Roman Reef.*

the Roman Reef, aquanauts Raymond Kientzy and André Portelatine moved from *Starfish* to the *Little House*, which was anchored at a depth of 26 metres. From here they could make long dives to depths of 40 or 50 metres and even make short descents to the 90- and 100-metre levels.

The first business trip to *Little House* was limited to a week. It was necessary to check whether nitrogen narcosis from the air they breathed was a potential problem for the aquanauts. Living conditions at this depth turned out to be more severe than at the ten-metre level of *Starfish*. Daylight could hardly be seen and the days seemed grey and dull. In addition, the living rooms of *Little House* were less comfortable than in *Starfish*. The main error was the lack of air conditioners. It had been thought that, at this depth, it would be cool enough without them, but the aquanauts suffered from the heat and humidity.

The *Précontinent 2* experiment was coming to an end when, one day, Soviet oceanographers from the research ship *K. Boldyrev*, invited by Cousteau, visited *Starfish*. When the ship with the oceanographers from the USSR arrived at Shab Rumi, the scientists were told that Jacques-Yves Cousteau and his

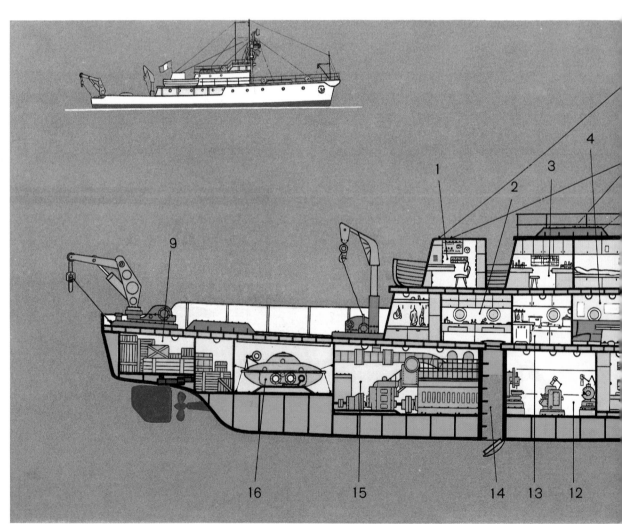

wife Simone were waiting for them under the sea.

Escorted by a group of French skin divers, the Soviet oceanographers made their descent. The undersea village made a vivid impression on the guests. Through valves in the roof of the continuously ventilated *Starfish*, excess air was being discharged. A sweeping train of bubbles extended to the surface. It looked like smoke coming out of a chimney. After climbing two or three steps up the ladder, the guests removed their face masks and entered the lobby. The hosts acquainted them with the furnishings of the underwater dwelling and told

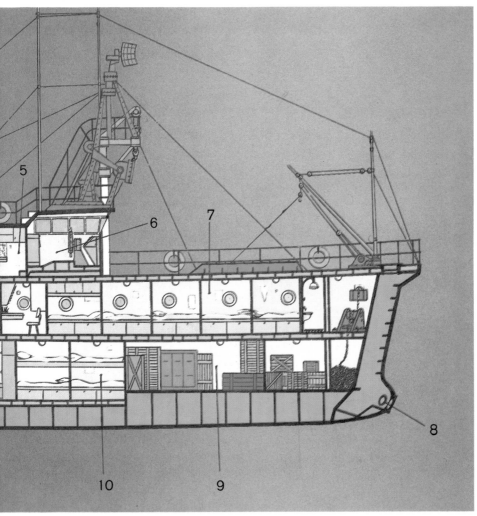

The name Calypso *is an honoured one in the history of oceanography. During its many years of service it has covered hundreds of thousands of kilometres and sailed in nearly every sea from the Arctic to the Antarctic.*

The crew of the Calypso gather on deck in the warm sunshine, always on the lookout for anything unusual or interesting at sea.

them about their investigations on the reefs.

At the end of their visit, the guests from the USSR were able to extend the circle of their acquaintances when they met the well-bred barracuda Julie not far from *Starfish*. Meantime, the tour of duty of *Little House's* crew came to an end and Cousteau gave the order to begin decompression.

Two more days went by and *Starfish* was also abandoned. On board the two ships a long-awaited meeting of the two crews took place: the dwellers of the undersea stations and the men who worked unceasingly at the surface. The aquanauts noticed how pinched and thin their companions had become. But everyone was satisfied and happy: the experiment was a success!

Précontinent 3

The underwater expedition in the Red Sea was not only a most trying ordeal that tested the endurance and professional training of its participants, nor simply an outstanding psychological and physiological experiment. It was also a critical examination of the engineering design concept involved. Therefore, all concerned in the events that took place at Shab Rumi had every reason to be proud of their exploits.

According to Cousteau, the most vital achievement of the Red Sea epic was the breath-taking realization that the sea could become a habitat of mankind.

Soon after he returned home Cousteau began to prepare a new expedition, hoping to establish an inhabited station at a depth of at least 100 metres.

The first trials were carried out, as before, with animals—goats and sheep. The four-legged aquanauts lived for about two weeks in a pressure chamber at a "depth" of 200 metres. Then they were replaced by specialists in underwater physiology, Professor Jacques Chouteau and Dr Charles Aquadro. They were to try for themselves all that awaited aquanauts isolated from the outside world under 120 metres of water at a pressure of 13 atmospheres.

The scientists spent many days within the walls of their crowded armoured-steel dwelling. There was no time to be bored, however. They made careful physical examinations of each other, and studied numerous electrocardiograms and encephalograms, registering their heart and brain activity.

Meanwhile, a new underwater station, *Précontinent 3*, was being built. It was an immense spherical buoy resting on a mount with telescopic legs. Crowds of tourists, strolling along the gaily decorated promenade in Monaco, racked their brains, wondering what this huge globe floating in the harbour, checkered like a chessboard, could be.

As well as Philippe Cousteau, responsible for underwater filming and photography, there were five more aquanauts in the crew headed by André Laban. So excited and happy that he could not conceal his feelings was the elder Cousteau too, supervising the final preparations for settling the new undersea dwelling.

The hour of departure came late at night. The steel sphere with its crew on board, towed as usual by two ships, the *Calypso* and the *Espadon*, left Monaco and headed for Cape Ferrat. Here, at the lighthouse, the shore station of the expedition was located. The aquanauts made their home two cables' length away, at a depth of 110 metres. In the early hours of September 25, 1965, all four legs of *Précontinent 3* smoothly came to rest on the sea floor.

Living conditions under water at Cape Ferrat turned out to be considerably harsher than in the shallows of the Red Sea reef. There was no hint of the fascinating interplay of light of Shab Rumi. Only a dingy blue haze in the daytime and impenetrable darkness at night. And, in addition, icy cold water. But it was always warm and cosy inside the dwelling. The aquanauts did not suffer from stuffiness or excess humidity as was the case in *Little House*.

Owing to the low conductivity of helium, Chouteau and Aquadro were bitterly cold during their on-shore experiment even when the temperature was 25°C. For this reason, a temperature of at least 30°C. was considered normal on board the station.

Diving with an aqualung in the Antarctic's Sea of Cosmonauts. The weather is favourable for a descent and Soviet biologists prepare to dive through a crack in the ice.

Special attention was given to medical and physiological research. All possible reflexes were tested; co-ordination of movement was regularly checked as were body temperature, heart activity, depth and frequency of breathing, and the pressure and composition of the blood. Not even the slightest

One of the miniature SP500 submarines encounters an underwater swimmer. There are two one-seater submarines of this kind in Cousteau's fleet capable of descending to 500–600 metres.

This species of triggerfish is one of the most spectacular of its family. It is an inhabitant of coral reefs in Pacific and Indian waters.

A group of soldierfish. They have sharp scales and bony, spiny heads. Their eyes are large, which suggests their nocturnal habit, as they usually hide in rock crevices during the day.

These ophiuroids, or brittle stars, have long flexible arms.

The Abudefduf, or sergeant-major fish, a flattened, blunt-headed fish with big eyes and tiny mouth, is usually found in shallow places among rocks.

physiological deviations were found.

Oceanographical research played a more modest part in the programme of *Précontinent 3* than in the previous expeditions. Microcurrents were studied and the generation of eddies in the bottom layer. Many samples of bottom sediments were collected. In the vicinity of the underwater house was a field of seaweed illuminated with searchlights. Without doubt, the bright light had a good effect on plant life. The development of phytoplankton was also accelerated by the artificial lighting.

The most memorable event in the underwater life at Cape Ferrat was the erection and testing of an oil-drilling rig. Petroleum engineers, watching this work on television, were delighted with the way the aquanauts performed all the necessary operations quickly and accurately. They worked so easily and with such skill that oil drilling might almost have been their normal occupation. The aquanauts freely manipulated 200-kilogram steel pipes without particular effort. They made ready the seals of the oil well and cleaned all the valves. The pressure of the oil in this "well" was imitated by means of compressed air.

In their trips to the open sea, the aquanauts could no longer employ ordinary aqualung equipment. The stores of air in the cylinders would not last long at such a depth. Consequently, they had to use hose-type apparatus for breathing. Naturally, the hoses appreciably reduced both the radius and the freedom of action of the aquanauts. But, to make up for this, they could stay under water as long as required to do their work.

The aquanauts spent three weeks at the sea bottom in their spherical hotel. On October 14, in the day-

Two underwater swimmers from Cousteau's team encounter a sea lion. This animal rivals the dolphin in its friendliness towards man.

time, after being freed from its ballast, the station *Précontinent 3* rose rapidly to the surface in only three minutes. Many people thronged the shore at Cape Ferrat. They had come to see the now famous checkered dwelling of the aquanauts. The meeting with the crew was delayed, however, for three and a half days. Only after 84 hours of decompression could the aquanauts come out of their globe in which they had lived for almost a month and again set foot on the soil of Monaco.

In answering the host of questions fired at him by newspaper reporters, Jacques-Yves Cousteau said, "In the first place we wanted to determine without doubt whether industrial work could be done at a depth of over 100 metres. Secondly, we wanted to reveal the capacity of the aquanauts, living in a synthetic atmosphere, for physical and mental activity. In both cases we obtained positive answers. *Précontinent 3* strengthened our belief in the necessity for undersea dwellings. But they should be independent of the surface to the maximum extent. This expedition clearly proves that people living at a great depth fully retain all their faculties. Our crew, as you know, was the first to live under such conditions. Consequently, the strain on their nerves was especially high. Nevertheless, each day they performed many strenuous tasks."

A nylon house

In June 1964 off the small isle of Berry in the Bahamas, the American explorer Edwin Link and his two assistants, the aquanauts Robert Stenuit and Jon Lindberg, had surpassed the 100-metre depth in the open sea for the first time, more than a year before their French colleagues. The new underwater habitation designed by Link, as

The sea horse clings by its tail to a sea fan coral. It can be found all over the world, mostly in shallow waters.

distinct from *Diogenes* and *Starfish* and other undersea stations, was made of a light, film-thin rubberized nylon. When filled with compressed air, the pneumatic house acquired its proper shape.

Not long before the decisive descent, the nylon house was tested at a depth of 20 metres. All the apparatus, both under water and on the surface, operated faultlessly. The only drawback was the somewhat crowded living quarters: only two metres in length and approximately the same in breadth and height. The furnishings were simple: a double bed, a tiny table, an electric air heater, a lamp and a refrigerator containing food. But there were a great many instruments controlling the composition of the air, the temperature and the pressure. There was also a telephone, a television camera and a telegraph for sending messages by Morse code.

After a short stay under water, the aquanauts returned to the ship. Meanwhile their house was lowered

to the extreme depth of 132 metres.

On June 30 at 9.30 a.m. the aquanauts entered the gondola which had served as a dwelling for Robert Stenuit two years previously and began their descent into the open sea. This took three hours and fifteen minutes.

The aquanauts left their diving bell and sent a message in Morse code to the surface to announce their safe arrival. Soon, however, they had to endure many unpleasant moments. The air purifier was out of order and the carbon dioxide content in the air began to increase menacingly. It became more and more difficult to breathe. A leaden burden pressed on their heads, dull heartbeats resounded throughout their bodies. Fearful of losing consciousness, the aquanauts put on their masks, sent the purifier to the surface for repair and returned to the diving bell.

By evening the air purifier had been repaired and sent back again under the sea. Meanwhile, it had become completely dark under water. Tired after the events of the day, Lindberg went to bed and Stenuit took the first night watch.

Next morning both men went out into the sea and began to test the new aqualungs designed for working at great depths. Close at their heels, keeping pace with them swam a huge sea perch. They found that life was abundant at such considerable depths: sponges, starfish, and a wide variety of fish and shrimps. Even though the aquanauts were dressed in special warm suits, both were bitterly cold because they spent so much time in the water. It was cold in the undersea dwelling too. To try to keep warm, Stenuit and Lindberg pulled on three woollen sweaters each.

At two o'clock in the afternoon of the following day the aquanauts were informed that the programme had been completed, and the crew could return to the surface. The order to get ready for evacuation was heard without any special enthusiasm.

"Before this, we would have agreed without hesitation. But now, when everything is going fine, we feel that we could spend several more days here—up to the weekend," wrote Robert Stenuit in his diary. But orders are orders, and Stenuit and Lindberg returned to the surface.

The house on a volcano

George Bond, who discovered the "saturation effect" and first realized the feasibility of living under water, was overtaken in his experiments by Jacques-Yves Cousteau and Bond's compatriot, Edwin Link.

Having in mind the time when man would be able to live and work in the depths of the sea and feel himself entirely at home there,

Bond planned an extensive programme of laboratory investigation. This work, that went under the name Genesis Programme in the history of aquanautics, took five years to complete. In the beginning, as usual, the heroes of the experiments were animals: white rats, rabbits, monkeys and goats. Shore experiments ended with a test in which three collaborators of Bond's spent 12 hours in a pressure chamber. Their return from a "depth" of 61 metres took 27 hours.

At dawn on July 20, 1964, Bond's associates—this time there were four—arrived at their new living quarters. In the Atlantic, 41 kilometres southwest of the Bermudas, on the peak of a dormant undersea volcano was a house resembling

Overleaf:
Précontinent 3 submerged at a depth of 110 metres off Cape Ferrat. In the daytime there is a blue haze of light but at night only pitch-black darkness.

The giant Kamchatka crab is caught in great numbers in northern Pacific waters.

Profile of the blue shark. The gills through which water is continuously filtered are prominently seen.

The underwater
observatory Sea
Lab 3. Tests on this
station were
discontinued after
the death of one of
the aquanauts.

Diogenes but much more spacious. Moreover, its design had to be more complex since the depth was 58.5 metres.

The crew, made up of experienced skin divers, Lester Anderson, Robert Bart and Sanders Manning, was headed by Lieutenant Robert Thomson of the Medical Corps. The fifth man was to have been Malcolm Scott Carpenter, one of the first American astronauts. But just before the descent, by unlucky chance, he broke his arm and the expedition Sea Lab 1 began without him.

As before in the shore investigations, the focus of attention was primarily the aquanauts themselves: their behaviour and how they felt within the walls of the underwater dwelling and outside, their capacity for critically sizing up a situation, for quickly making the proper decisions and for performing the planned work.

It was intended that the aquanauts should stay at the bottom for a complete month. But on the eleventh day a report from the weather forecasters contained warning of a coming storm. This was a threatening danger signal, for the existence of Sea Lab 1 depended wholly on the surface base. The storm could break loose the ship from its anchor and damage the hoses and cables connecting the station with the undersea dwelling. It was decided to start bringing up the aquanauts at once. Despite the raging storm which curtailed the experiment, George Bond was satisfied. The capacity of man to live under water in a hostile medium was brilliantly confirmed yet again.

At the edge of the precipice

A year later Sea Lab 1 was superseded by a new undersea residence. It was built on the basis of experience gained in the Bermuda experiment. Having ballast tanks like a submarine, Sea Lab 2 could surface and submerge independently. Electric power and fresh water were supplied from a shore base and all the required gases for preparing heliox were on board the station.

Ten American aquanauts, headed this time by M. Scott Carpenter, moved into the spacious residential compartment of Sea Lab 2, anchored at a depth of 61 metres. Diving down to the entrance hatch of the dwelling, Carpenter was the first to cross its threshold. It was warm and comfortable in the rooms of Sea Lab 2, and the aquanauts removed their face masks, aqualungs and rubber suits. But outside the walls conditions were entirely different and far from subtropical.

"For Sea Lab 2, I chose the blackest, coldest and most terrible water that could be found near the shores of America: the edge of the Scripps Submarine Canyon in California," said George Bond.

To keep warm, the aquanauts of Sea Lab 2, like their French colleagues, wore special rubber suits with electric heating facilities. Unfortunately, these suits were not very reliable. But without them it was impossible to go out into the open sea at great depths. The aquanauts complained that with ordinary rubber suits it was only possible to keep warm by working vigorously.

Life under the sea was also

Aquanauts from the Sea Lab 3 crew. On the left is Malcolm Scott Carpenter, the American astronaut and deep sea explorer.

An underwater installation which powers electrical instruments used by the aquanauts.

which was entered almost at once by several fish of their own accord. Certain other specimens of sea fauna that especially interested the scientists were caught with nets and made to live within the cage.

The underwater dwellers were quite safe from sharks: a trained dolphin called Taffy became their defender and guarded them against these ferocious man-eaters. Taffy also had other, no less responsible, chores to do. He was trained to take a lifeline to aquanauts who lost their way in the dusky waters of Scripps Canyon. He could immediately be summoned by whistling. Taffy was also a conscientious postman. Moving between the undersea station and the shore base, he delivered a number of letters, newspapers and magazines to the crew of Sea Lab 2.

With the same diligence he delivered heavy packages with spare parts, tools and food supplies. For these services Taffy was elected an honorary member of the Association of Postal Workers of the USA.

Like *Précontinent 3*, the Sea Lab 2 expedition was an important step forward in the conquest of the continental shelf.

Ichthyander of Taurica

In the western part of the Crimean peninsula, in ancient geography the land of Chersonesus Taurica, near steppes baked in the fiery sun and dried by burning winds, a narrow strip of land stretches far into the sea. This is Cape Tarkhankut. Translated from the Turkish this means "Devil's Corner". Here was a tent settlement of aquanauts, sportsmen from the Ukranian city of Donetsk. Members of the skin-diving club Ichthyander they were the first in the USSR to build a lodging under water, which they moved into during 1966.

spoiled to some extent by the tricks played by the Yankee gas, as helium is sometimes called. Not only people suffered from it but instruments as well. Among the first victims were the television cameras. Penetrating inside them, the heliox greatly impaired the degree of contrast of the transmitted pictures.

Fifteen days after the beginning of the expedition, a second group of aquanauts moved into the dwelling. This was again headed by Carpenter. During his long undersea stay, Carpenter was able to radio his colleagues in space, the astronauts L. Gordon Cooper Jr. and Charles Conrad Jr., in their flight on the orbital spaceship *Gemini 5*.

Members of the Sea Lab 2 crews included experts in various fields. The comprehensive programme of the expedition covered over 40 different assignments. An important place was allotted to construction on the sea floor. Like their French colleagues, the men were engaged in setting up and operating an underwater oil-drilling test rig. Marine geologists and hydrologists studied the ocean floor and undersea currents. The biologists erected a huge wire aquarium near the dwelling

This debut by the Donetsk skin divers was a great success. A year later, their expedition was located in the vicinity of Sevastopol, near the rocky cliffs of Laspi Bay. Hoses and cables twisted snakelike among the chaotic heaps of enormous rocks. They ran to the sea and disappeared under water. From the shore control desk, perched on the slope of the cliff, you could see the whole bay and Cape Sarych, the southernmost point of the Crimean peninsula.

Pteropterus radiatus *is one of the strangest looking and most dangerous of fishes. A prick from one of its poisonous thorns can cause a painful wound.*

Dolphins seem happy to be close to man. In captivity they are easily trained to perform for the public's delight.

The sea squirt, or ascidian, anchors itself firmly to the sea bed. Through an opening or "mouth" on the top of its body the animal draws in water and feeds on the plankton it contains.

Some species of starfish, are predatory and aggressive, even attacking their own kind.

Butterflyfishes form a large group which inhabits tropical and temperate waters. They are noted for their brilliant colouring.

The moray eel, with its sharply pointed teeth, can be dangerous and aggressive. The largest species grows to three metres in length.

Predominant among the aquanauts were two professions—mining engineers and doctors. For this reason medical and physiological research and practical mining and geological investigations at the sea floor were the basis for all the Ichthyander programmes.

The aquanauts from Donetsk settled at a depth of 12 metres. Their new home had four spacious compartments: a comfortable bedroom, galley, shower-room and a deck-cabin for controls and communications which was also the laboratory. The first group to go under the sea consisted of five aquanauts headed by Alexander Khaes. The day after they moved in, an alarm was raised. The pressure suddenly dropped and water rushed into the undersea dwelling, its level rising rapidly. An SOS signal was given in the camp. Fortunately, the shore crew quickly located and eliminated the trouble. A week later a second crew went on duty.

Later, the aquanauts of the Donetsk skin-diving club conducted several more experiments on the floor of Laspi Bay which made a substantial contribution to the history of aquanautics. Of special interest was the testing of a new, lightweight helmet-type diving suit. Made of multilayer elastic rubberized cloth, this suit keeps a diver warm in cold water. Aquanaut Igor Motsebeker spent 26 hours and 15 minutes in the suit under water. Even longer, almost 37 hours, was the underwater watch of Sergei Khatset.

The aquanauts communicated with the surface, ate meals, made notes, listened to the radio, read the newspapers enclosed between two sheets of transparent plastic, slept, all without removing their diving suits. Four hours of sleep in hydraulic weightlessness proved sufficient.

The house on the slope

Undersea explorations have an immediate appeal and news of them, published from time to time, is read with no less interest than adventure stories. In this respect, the activities of the Underwater Research Laboratory of the Leningrad Hydrometeorological Institute are no exception. The undersea researchers of this institute were the first in the USSR to put lightweight diving gear into regular service for scientific purposes.

An expedition of the institute worked for three years in the underwater oil fields of the Caspian Sea. The aquanauts were engaged in examining underwater structures: the legs of the trestles and drilling "islands" and the well heads. When their programme of investigations in the Caspian Sea had been completed, the aquanauts from Leningrad decided to settle down in the Caucasus on the floor of the Bay of Sukhumi. They named their undersea dwelling *Sadko* after the Novgorod singer who accompanied himself on the psaltery. According to legend, Sadko had been to the sea bottom where he was presented at the court of the Sea King, returning afterwards safe and sound to dry land.

Sadko resembled Cousteau's checkered globe *Précontinent 3* to some extent, but was somewhat smaller. It was held to the sea floor by a ballasting anchor. A wire rope stretched from *Sadko* to the anchor. If the rope was paid out, *Sadko* rose, if the winch was reversed, *Sadko* slowly sank. This was its advantage over other undersea dwellings, enabling it to be used at this particular location where the bottom slopes at an angle of 40°.

Sadko made its first descent in 1966. Its inhabitants, first at a depth

Jellyfish, umbrella-like coelenterates with drooping oral stalks and tentacles, are a common sight at sea. Small fishes often seek protection within the cupolas of big jellyfish.

The base of this underwater mast is fitted with sensors for making hydro-optical observations, such as registering the degree of illumination in the depths of the sea.

thermocline can be seen because it is the habitat of plankton and drifting organisms such as jellyfish which are caught between the dense, cold, deeper layers and the less dense, more saline and warmer upper water. Dead seaweed is also found there. The aquanauts employed current indicators. This made the dynamics of the thermocline visible. They observed a peculiar phenomenon: at the upper and lower boundaries of a layer only a metre and a half thick the water flowed in opposite directions.

Chernomor's *first summer*

Chernomor was the name of the powerful magician in Pushkin's fairy tale *Ruslan and Ludmila*. It was chosen by the research workers of the Oceanographic Institute of the USSR Academy of Sciences as the name of their undersea dwelling. Like *Sadko*, *Chernomor* was located at the Caucasian seashore. Near the city of Gelenjik, it was anchored at the bottom of an inlet with the poetic name of Golubaya (Sky-Blue) Bay. Standing on short thick legs it resembled some kind of mythical beast with its tail raised and a curved beak.

Chernomor welcomed its first settlers in the summer of 1968. On August 9, a scientific crew headed by the oceanographer Pavel Kaplin moved in at a depth of 14 metres. Kaplin was one of the first in the world to dive with an aqualung to a depth of about 100 metres. But, as one of the designers of *Chernomor* said, "It was not our purpose to reach record depths and stay as long as possible under the sea. Our chief aim was to ensure the absolute safety of the crew and to provide maximum comfort under water. Scientists should be able to work just as easily on the sea floor as in their dry-land laboratories. There-

The underwater observatory of the Oceanographic Institute of the USSR Academy of Sciences is named after the fairytale character Chernomor, master of the Black Sea.

Previous page: Diving experts of the Comex company at work on an underwater oil well. Their diving bell is seen on the right.

of 30 and then 40 metres, were two rabbits and a dog. After they had spent two weeks under water, the watch was taken by eight crews of aquanauts who took turns to live in the dwelling.

A year later a new residence, *Sadko 2*, was lowered into the Bay of Sukhumi. Its submergence was a grand occasion. According to marine tradition, a bottle of champagne was broken against the wall of the station which now had two spherical storeys. The crew of *Sadko 2* consisted of Veniamin Merlin, an engineer, and the oceanographer Nikolai Nemtsev. The first stop was at the 11-metre level. When they felt at home in the new dwelling, the aquanauts were lowered with it to a depth of 25 metres.

Interesting observations were made of the thermocline. This is a particular layer of water in the sea where the temperature, salinity and density change abruptly, the change being readily felt by a swimmer who suddenly passes from a zone of warm water into a cold one. The

fore, we planned the underwater dwelling in the best possible way for research purposes."

The scientific programme of the *Chernomor* expedition was comprehensive and highly varied. Scientists in various fields, hydrophysicists, hydro-opticians, geologists and biologists worked aboard the undersea laboratory. Five crews, united by their scientific interests, lived in turn at the bottom of Sky-Blue Bay.

Chernomor's furnishings resembled those of a ship's stateroom. Bunks were arranged, two-high, along the walls. There was a bookshelf, a table and a control desk. Only the red emergency aqualung at the head of each bunk, and the equipment for working under water, neatly piled at the entrance hatch, were reminders of the sea hidden behind the walls of this comfortable dwelling.

When *Chernomor's* inhabitants swam late in the evening or at night among the shaggy seaweeds growing so abundantly in the offshore zone of the Black Sea, the thick growths would begin to shine brightly. Then the aquanauts began to radiate light too. Their hands shone, as did their feet in swim fins and their hair. Even the hoses of the aqualungs gave off light. The cause of the mysterious bioluminescence was a tiny sea animal, the flagellate, which lives in the seaweed jungles. In calm weather, the seaweed is usually in darkness. But upon even slight ripples, the flagellates are stirred and the murky gloom is lit up by thousands of minute lights.

The marine geologists and hydrophysicists had much to do. To conduct observations, the floor of Sky-Blue Bay was divided into 12-metre squares, or test fields. The furthermost corner was 350 metres from the dwelling.

The scientists decided to tint the oil of the test field with luminophors. This was a tedious job, but it enabled them to observe visually the drift of the sand, the accumulation of bottom sediments, and the undersea currents.

Thirty days under water passed by almost imperceptibly. The *Chernomor* inhabitants completed their programme of research and the last crew prepared to leave the undersea dwelling. Decompression procedure was not difficult. It was only necessary to seal the entrance hatch tightly and the underwater laboratory became a decompression chamber. The pressure in *Chernomor* was then gradually decreased until it was normal.

Hymenocerca elegans are exotic-looking shrimps which live in tropical seas.

Brittle stars can detach one or more arms in self-defence, or even break up in two. Each piece shed is capable of regeneration into a new creature.

When a sea urchin is pursued by a starfish even its needles do not prevent it being caught and swallowed.

CHAPTER 4

SECOND GENERATION

After the success of the expedition *Précontinent 2* in the Red Sea, Jacques-Yves Cousteau justly claimed that an ancient dream of mankind had been realized. Man had lived under water and could, when he liked, leave his undersea dwellings and return to them without rising to the surface. Now it was known that if they used the proper gas mixtures for breathing and followed the pertinent rules, people could live for long periods at the sea bottom and gradually develop the extensive ranges of the undersea plateau.

The subsequent course of events confirmed Cousteau's optimistic prediction. Later undersea expeditions, in which aquanauts from France, the United States, the USSR, Great Britain and other countries took part, left no doubt as to the validity of the ideas, once thought far-fetched, expounded by pioneers of undersea "house-building".

For the most part, the first underwater dwellings were firmly tied to their attendant ships, or shore bases which supplied the aquanauts with everything required for life on the sea floor: electric power, fresh water, hot food and containers with compressed air. But it was frequently this link with the surface that proved to be the Achilles' heel of the undersea houses.

A "second generation" of underwater dwellings began to appear at the beginning of the 1970s. They were more reliable and more comfortable than their forerunners and were remarkably self-contained.

In undersea orbit

The idea conceived by officers of the National Aeronautics and Space Administration (NASA) at first surprised even experts in aquanautics. NASA proposed that the undersea dwellings be used as a training device for simulating conditions encountered on a prolonged space expedition and of life on a base established on some other planet. This idea, which might be thought absurd, was actually of great intellectual depth. In no place on our planet can we create for a group of investigators such a situation of maximum isolation and estrangement from their habitual earthly life as the sea depths. The life and safety of aquanauts, like astronauts, depend upon the reliability of their shelter and upon the equipment that fills it. Both breathe artificial air and, when they leave their sanctuary, employ protective clothing with a self-contained life support system. Finally, aquanauts experience hydraulic weightlessness which closely resembles space weightlessness.

A new undersea station was

On the threshold of the unknown. Each aquanaut has to learn for himself the dangers and delights of the underwater world.

Opposite: As the aquanauts descend, the underwater world becomes grim and dark. Daylight penetrates for only a short distance, as it is absorbed and dispersed by the water.

This species of ctenophore, equipped with four pairs of vividly fluorescent rowing plates, is frequently encountered in polar seas.

designed named *Tektite*. (In general use, the word "tektite" means a mysterious glassy body found now and then on earth. It is supposed to be of meteoric origin). The dwelling resembled an orbital station. It consisted of two vertically oriented capsules, held on a common mount, with telescopic legs and joined together by a short passage tunnel.

In one of these towers the diving compartment was on the first floor and the machine room and life support system were on the second. In the other tower there was a four-bunk stateroom underneath and the laboratory with control and communications compartment up-

The undersea observatory was set up at a depth of only 15 metres. The work day of the crew began at eight o'clock in the morning. After breakfast the aquanauts went out into the sea in pairs, returning from time to time to the house to replenish their air supply in the aqualungs.

They finished work in the early evening. Removing their aqualungs and rubber suits, they took a hot shower to warm up and put on other clothes. Then, after checking all the major apparatus of the undersea dwelling—and this was done daily—they sat down to supper.

After supper they sat down to paper work, bringing their notes up to date. This was followed by cursory medical examinations performed on one another. Then they reported to the surface that everything was in order and checked the emergency system of the dwelling. Finally, near midnight, they went to bed. At first the inhabitants of *Tektite* would leave somebody on duty at night. After two weeks, however, even this safety measure was abandoned. Thus they lived, day after day, without incident.

The splendid coral reefs of the Caribbean, surrounding the station on all sides, proved to be excellent natural laboratories for conducting highly interesting biological research. What were the coral reefs like near the location of *Tektite?* What were they inhabited by? Were the reefs growing or diminishing? What did the coral polyps feed on? How did they multiply and how quickly did they grow?

Above right: The spotted moray is the leopard of the sea depths. Despite its ferocious appearance this specimen was on good terms with Calypso's crew, who treated it to pieces of meat.

stairs. Semispherical plexiglas portholes in the walls of the compartments provided a panoramic view.

The new undersea observatory was installed on the floor of Great Lameshur Bay off the island of St John in the Virgin Islands in the Caribbean Sea. After carefully considering all the candidates for the crew of *Tektite*, the men chosen were oceanographers Richard A. Waller and Conrad V. W. Mahnken, geologist H. Edward Clifton and marine biologist John van Derwalker.

The station's inhabitants also studied the life of fish, crustaceans and shellfish. Experiments with them proved fascinating. Some of the animals were tagged with tiny supersonic beacons the signals from which could be detected within a six-kilometre radius. Others were

supplied with fluorescent tags. These sonar and light signals revealed much of interest about the life and habits of the inhabitants of the sea.

In the course of time, when they became accustomed to undersea life, the aquanauts would swim farther away from their house. Excursions at night also became common. After living two months under water, which was a record, the crew of the hydraulic space-training observatory came to the surface with a record rate of decompression. It took only nineteen hours. The gas mixture used by the aquanauts was

Unlike reef-forming corals, sea anemones have no limestone skeletons and disappear after death leaving virtually no trace of their existence.

also unusual. It is known as nitrox and contains eight per cent oxygen and 92 per cent nitrogen. The *Tektite* expedition proved that at a depth of 15 metres such a gas "cocktail" is quite harmless.

Richard Waller, the commander of the expedition, was convinced that people who were going to live on the ocean floor could do a great deal. The quality of the work carried out by the aquanauts would doubtless have improved if their stay under the sea had been prolonged. After two months of life on the sea floor, all the aquanauts felt amazingly well and claimed they could have stayed down longer.

Tektite II

Soon after the successful completion of the *Tektite* expedition, NASA, convinced that they were pursuing the proper course, decided to continue underwater research using the same station at the same depth. The scope of the new expedition was unprecedented. It lasted for eight months, from April to November 1970. During that time, ten underwater crews took turns living on the sea floor.

For the first time in history, a crew made up entirely of women and headed by the geologist Sylvia Erle, lived at the bottom and spent two weeks under water. One of the *Tektite II* crews was international, consisting of men from the United States, France, West Germany, Australia and Brazil. In all a total of 53 aquanauts took part.

The aquanauts in expeditions *Tektite I* and *Tektite II* were engaged in research and were themselves objects of painstaking observation by physicians, psychologists and other specialists.

Before the start of their underwater ordeal, the aquanauts filled out questionnaires, at the request of the psychologists, answering a great many questions of every possible kind. Special attention was given to questions about the life they led in childhood and adolescence. Computerized processing of all the data enabled certain conclusions to be drawn about the probable behaviour of the aquanauts according to the conditions of their early life and upbringing. For example, the most productive work was done by crew members who had grown up in comparatively small communities with firm and stable modes of life. Such people, it was found, have a much more developed sense of co-operation with others.

The residents of large cities were, to a greater or lesser extent, more likely to be individualists.

Also of significance were the skills and habits acquired in childhood. Aquanauts who were the first-born in their families were more broad-minded and better informed on the latest achievements in science and engineering, but they reacted more keenly to the shortcomings of their colleagues. Over half the members of the Tektite crews turned out to be the first-born in their families.

Sadko 3

At about the same time as the *Tektite* expeditions, on the other side of the Atlantic in the Black Sea, the Soviet observatory *Sadko 3* was put into undersea "orbit". Externally it resembled a three-stage space rocket. The lowest "stage" was the diving compartment, the middle stage, the residential area and galley, and the top stage the area for scientific apparatus. The resemblance of *Sadko 3* to a rocket was accentuated by the oblong capsules that were attached outside the diving compartment and which looked like rocket engines. They were actually water tanks, the decompression chamber, lavatory and storerooms of the underwater station, with an entrance from inside the diving compartment.

The unusual design of the new observatory was due, as before, to the conditions on the floor of the Bay of Sukhumi into whose slope the 30-tonne anchor of *Sadko 3* was moored. The chain running up from the anchor firmly held the undersea dwelling perched at the edge of the escarpment. If the ballast tanks were blown and the anchor raised, *Sadko 3* surfaced like a submarine.

The first to enter *Sadko 3* were the testing crew consisting of Vsevolod

Djus, designer of the station itself, Alexander Monkevich, designer of the electronic equipment, and John Rumyantsev, an experienced diver. During their first hours under the sea, the aquanauts heard the TASS report on the launching of the

The flame scallop belongs to a group of molluscs called bivalves. Its mantle is edged with fringed sensory tentacles.

A school of sea perches grazing in the coastal waters of the Sea of Japan. They are carnivorous fishes and dart from places of hiding to attack their victims.

Polychaetes, or sea worms, can be found in all waters and at all depths. These animals are frequently so great in number on the sea bottom that they form colourful living carpets.

73

The octopus is a great expert in camouflage.

This bare-gilled mollusc is tree-like in form.

A starfish moves from stone to stone in search of food.

Underwater enemies — a sea urchin and two kinds of starfish.

Soviet spaceship *Soyuz 6* into orbit with cosmonauts Georgi Shonin and Valery Kubasov on board. Kubasov later was to become famous as a participant of the Soviet-American space experiment Soyuz-Apollo. Most remarkable was the fact that the spaceship was launched on the same day and in the same hour as *Sadko 3* was submerged.

The first testing crew carefully checked all the systems of the under-sea dwelling. When they had finished, at the end of the third day, the aquanauts were ordered to leave the station and to enter the diving bell which was to lift them to the surface. On the same day their places were taken by a scientific crew consisting of Commander Anatoly Ignatyev and the aquanauts Yevgeni Savchenko and Valentin Bezzabot-nov.

Sadko 3 was anchored at a depth of 25 metres. If necessary, it could be lowered to twice or three times that depth because it was built with a high safety factor. But the Leningrad aquanauts, like the *Tektite* crews, were not aiming at record descents. The expedition had entirely different assignments: to study the life and behaviour of sea fish, and to record and decipher their voices.

As recently as 20 or 30 years ago, the sea depths were regarded as a place of silence. Today we know that this is not so. The investigation of the acoustic capacity of fish began only in our time. By means of special instruments—hydrophones—we can now hear the sounds they make. For fish emit supersonic sounds that cannot be detected by the human ear. Thus *Sadko 3* recorded several solo and choral performances by fish and, at the same time, the sounds made by other sea animals.

But how can these recordings be

A chance aquanaut — the kitten Caissonka. A stowaway on board the underwater station Sadko 3, it would frequently find its own food. Perched at the edge of the entrance shaft, the kitten would catch fish attracted by the bright light.

interpreted? How can we come to know the "language and dialects" of underwater creatures? What do their sounds mean? Pleasure experienced from tasty food or threats aimed at an enemy? Perhaps they are signals of danger or appeals to kinsmen for aid and protection, or a family quarrel, or maybe the last dying cries of a victim? This is what the investigators of *Sadko 3* were to try to unravel.

The aquanauts built a special cage: a metal apron frame of cylindrical shape covered with a net. The cage was about ten metres in diameter. It was hung on the body of the underwater station, surrounding it from all sides like a lampshade. Only the entrance hatch of *Sadko 3* was left open.

The first to be put in the cage were the horse mackerel. Their loss of freedom was very difficult to bear even though all that fenced them off from their native element was an undulating, almost invisible cobweb-like net. The silvery fish restlessly dashed from side to side, seeking an escape from captivity, but not finding one. Even though they had plenty of food, some of them died soon after. Next to be put in the cage were croakers, sea bream, rays and spiny sharks, the only species of their family found in the Black Sea. Observing them, the aquanauts found that fish are especially talkative between nine and eleven o'clock in the evening.

To study the reaction of the fish to the "language" of their fellows, the

aquanauts recorded their sounds and then played them back through a loudspeaker, allowing first one and then another species of fish to give a "concert". As a result of many days of observation, the crew of *Sadko 3* accumulated extensive data on the way of life, habits and behaviour of Black Sea fish.

When the "language" of fish has been exhaustively studied and thoroughly understood, fishing fleets will probably lure schools of fish, rather than hunt for them, broadcasting fish "voices" recorded on magnetic tape as signals calling the fish of a definite species to gather in schools. This is an interesting development, one of great practical significance in studying the bio-acoustics of the sea.

Among other assignments given to the aquanauts was one calling for the testing of special dishes intended to feed cosmonauts on a space flight.

Chernomor 2

One of the most successful of the underwater research stations, *Chernomor 2*, began its long career in 1969 in difficult and stormy conditions. On the second morning after the test crew had taken up residence in the dwelling, Sky-Blue Bay foamed with muddy waves and the station, anchored at a trial depth of 12 metres, was severely shaken by the waves.

Next day there was no improvement in the weather. In fact, the meteorologists forecast an increase in the intensity of the storm. The first victim of the raging sea was a catamaran-type pontoon anchored near the underwater observatory. Its anchor broke loose during the fierce attacks of waves and wind and the catamaran was cast ashore with its sides broken in. Next, the same disaster befell the expedition's supply ship *Do-ob*. The undersea dwelling, however, withstood the storm and only minor repairs were required after it had abated. So the storm, if unwelcome, did confirm the reliability of the new undersea house and of its vital systems.

In building *Chernomor 2*, the hull of its forerunner *Chernomor 1* had been used, although everything else was new. Changed beyond recognition were not only the life support systems, but the external appearance of the station as well. A conning tower was built on the upper deck like that on a conventional submarine. Underneath the station were hydraulically controlled supporting legs made from the landing gear of an *Il-18* airliner. The hull was lengthened and platforms provided at the bow and stern with holders for huge gas cylinders firmly attached. The gas supply was increased 50 times, the supply of fresh water sixfold and the previous stor-

age batteries were replaced by ones with a capacity 100 times greater. In comparison with the first underwater station of the Oceanological Institute, *Chernomor 2* was heavier by 25 tonnes.

The air and power lines connecting *Chernomor 1* to the shore were the most vulnerable parts of the station. *Chernomor 2*, on the other hand, was practically self-contained. There were three channels of communication with the surface observers: radio, telephone and closed-circuit television. A duplicating radio buoy was on patrol at the surface. In a word, the new *Chernomor's* inhabitants were fully equipped to extend their stay in the world of hydrospace.

In July 1969, a team of powerful tractors and bulldozers towed the station to the water's edge. Then a 100-tonne crane picked it up and gently lowered it into the water. At

On board the underwater station Chernomor 2.
1. battery compartments;
2. crew's quarters;
3. work room;
4. diving compartment;
5. oxygen cylinders;
6. diving shaft;
7. ballast sections.

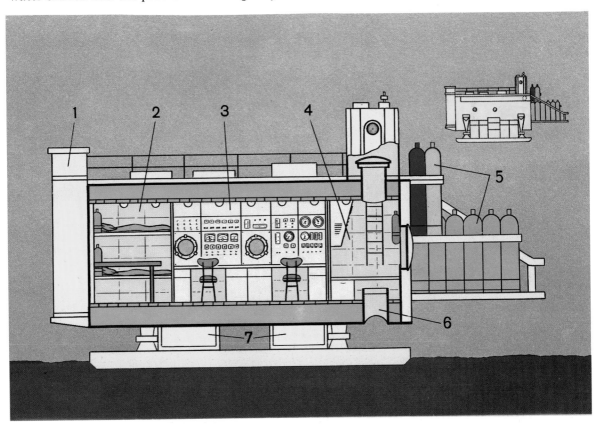

the beginning of August, the ship *Sestra* sailed into Sky-Blue Bay to take the place of the wrecked *Do-ob*. Everything was ready for the next round of tests. The command to submerge the station was given at noon on August 15.

Exactly two weeks later, the first scientific crew took its place in the station. It included aquanauts who were geologists, assigned to investigate the bottom deposits of Sky-Blue Bay. Substantial changes in the relief of the floor had been noted here and large accumulations of sand and gravel observed. In studying these deposits, they were found to be brought by powerful undersea currents which appear from time to time in the bay. In the beginning, of course, it was the waves, gradually demolishing the shore, that had created them.

The geologists were relieved by a

Chernomor 2 is about to be lowered by a powerful 100-tonne crane into the waters of Sky-Blue Bay.

Aquanauts from the Soviet underwater house Sprut *examine the seaweed-covered rocky reef on the floor of the Black Sea.*

crew of hydro-opticians. The aquanauts working in this field continued to investigate the luminance of the light in the sea depths. This research had been begun in the preceding year. It is no simple matter because the intensity of light under water continuously varies, seeming to pulsate. This is caused by changes in the weather and the roughness and depth of the sea.

The reflection of sunbeams moving on the sea bottom should give data on illumination at the surface of the sea and in its depths, the optical properties of the water, light conditions in the course of the day and night at various depths, the absorption of light and its range of propagation. Many aspects of this research were entirely new.

Biological proving grounds that made their appearance in the vicinity of the undersea station

were reefs made by hand out of chaotically piled concrete blocks, slabs and sets of steel shelves. Investigating the artificial reefs from time to time, the biologists were pleased to find that their population was increasing. Among the sea

The underwater station Aegir *seen here stationed at the pier of the Hawaiian Oceanographic Research Institute.*

Members of the crew of Chernomor *relax in the living quarters of their station.*

Opposite: Chernomor 2M *was designed to create living and working conditions suitable not only for professional divers but for those scientists who had mastered the art of aqualung diving.*

Marine biologists at work in their natural laboratory.

This species of dogfish is relatively small but is extremely aggressive. It attacks crustaceans, molluscs and other sea animals.

The sea urchin is always a danger to the unwary diver. Its long poisonous needles can pierce the skin and cause a painful wound.

Female octopus protecting her eggs laid in the depth of a stone niche.

fauna who found a haven there were horse mackerel, sea perch and ruffs, not to mention mussels and the bitterest enemy of mussels and oysters, the rapana. Marine biologists are of the opinion that such "dormitories" will be useful in the future in organizing fish farms in the sea.

The whole of the next undersea season was given over to research by medical personnel and physiologists. Work began in early spring and finished at the end of the summer. Along with the underwater station, the diving compression complex operated fulltime. Several dozens of aquanauts completed a course of training and tests were conducted in the two-compartment pressure chamber. The depth of these "dry dives" was gradually increased from 20 to 100 metres, with a change-over to heliox for breathing. Not all the aquanauts could conquer these depths and some were eliminated. Most, however, passed the tests easily and were ready to repeat the dives under natural conditions. For the time being, however, this was not part of the Oceanological Institute's plans.

By the summer of 1971, new improvements had been made in the station, now called *Chernomor 2M*, to make it fully self-contained. It was decided to start the new expedition without providing any kind of floating base at the surface. By the recommendations of the doctors and physiologists who drew up the programme for the expedition, *Chernomor 2M* was this time anchored at a depth of 31 metres, the maximum depth of Sky-Blue Bay.

In formulating his plans, Andrei Monin, Director of the Oceanological Institute, told the aquanauts that in his view scientists should concentrate on producing scientific data on all that takes place in the

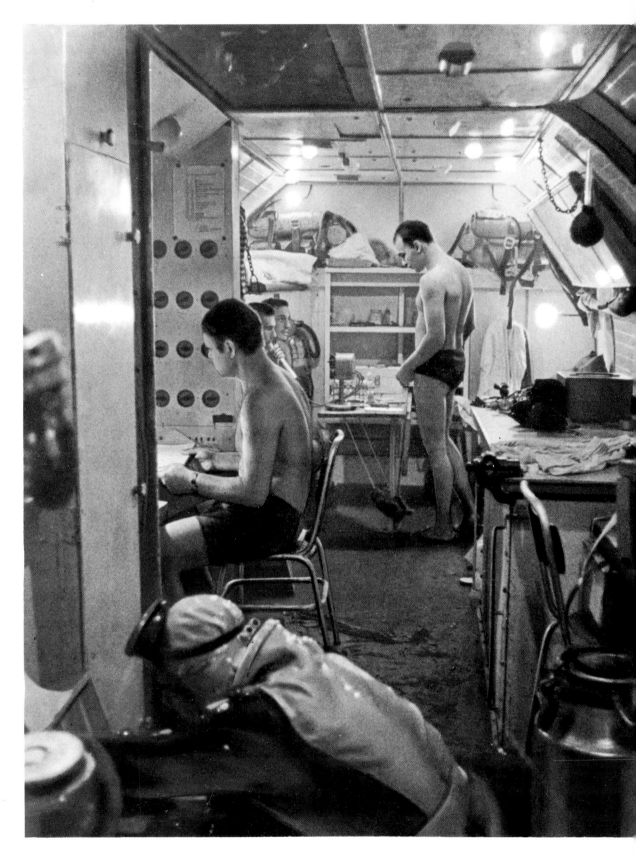

Divers have made many underwater discoveries of archaeological interest. Over this ship's wreck swarm a shoal of Abudefdufs, *or sergeant-major fishes.*

maintaining constant communication with the shore.

The aquanauts spent the summer of 1972 at a depth of 30 metres. When the test ended, a crew headed by Igor Sudarkin moved into the dwelling. The investigators lived under water for three weeks and were then replaced by a second scientific crew headed by Oleg Skalatsky, an engineer-physicist. In spite of the extremely severe conditions—a storm raged almost continuously at sea, visibility under water was very poor, and the roughness of the sea was apparent even on the bottom—Skalatsky and his comrades courageously carried on. They lived on the sea floor for 24 days.

The *Chernomor*, *Tektite* and other similar stations were part of extensive research programmes carried out in the Soviet Union, the United States and other countries. The aim of these expeditions was to convince society that man is capable not only of living under water for a long time in conditions of isolation, but also of conducting research and of extracting commercial minerals for industry's use.

The *Chernomor* expedition, planned to cover a period of several years, finally ended in well-deserved success for the undersea investigators of the Oceanological Institute of the USSR Academy of Sciences. Exceptionally valuable information had been obtained on life in the sea at depths up to 30 metres, experience had been gained in the organization of such expeditions, and a detachment of aquanauts with particular scientific skills had been recruited. Much had been learned by the designers of the undersea equipment, which was continually improved and renewed. Whereas the aquanauts of *Tektite* had proved the suitability of nitrox at a 15-metre depth, the crew of *Chernomor 2*

wave and breaker zone of the oceans and not seek to break records. So the station was returned to a depth of 15 metres, although the term of continuous duty by the aquanauts on the sea floor was prolonged to two months.

At first it seemed there was plenty of time and no reason whatever to hurry. But the aquanauts knew this impression to be deceptive, for the schedule each day was very rigorous. Morning and evening, sessions of medical and physiological research were conducted. During the day, and often in the evenings, work was carried out in the sea and in the compartments of the station. An around-the-clock watch was kept at the central desk, the nerve centre of the undersea station, tested it at twice that depth during many months of life on the floor of the Black Sea.

Next summer, *Chernomor 2M* was far away from Gelenjik, off the shores of Bulgaria. Professor A. S. Monin, director of the Oceanological Institute, thought that the undersea world in this new location would be richer, more interesting and more suitable for habitation than in the vicinity of Sky-Blue Bay. So the USSR joined Bulgarian specialists in a joint expedition. The debut of *Chernomor 2M* on the international scene was just as successful as on her own territory.

Under the Baltic Sea

Experts of NASA were not the only ones to have the idea of using an undersea dwelling to conduct "extra-terrestrial" investigations. Professor Kinn, director of the Heligoland Biological Institute and Professor Ruff, head of the Insitute of Aeronautical Medicine in the West German Centre for Air and Space Communications agreed to build an undersea laboratory for conducting joint research in hydro-space.

The first undersea dwelling in the Federal Republic of Germany, named BAH after the Heligoland Biological Institute, was distinguished by a number of innovations. It could be dismantled into separate units, making transportation easier, and had exceptionally good thermal insulation. The first submergence of the undersea dwelling took place in September 1968 in the Baltic Sea. The weather at sea was rough at that time and BAH held onto the sea floor for only two weeks. Notwithstanding the small amount of data obtained, Professor Kinn declared that he had no doubt about the importance of undersea research stations.

Soon afterwards, a new station, named *Heligoland*, was built in the Federal Republic of Germany. *Heligoland* combined the best features of many underwater dwellings that had been built previously. It was to serve under the rigorous conditions prevailing in the North and Baltic Seas, in regions with unstable, cold weather, with frequent high winds and strong underwater currents. To cope with these conditions, the

designers took special care to make the new laboratory self-contained.

One of the remarkable innovations of the expedition was its unique floating buoy *Füstchen*, named in honour of its inventor, the engineer Füst. This 16-tonne structure resembled the engine room of a small ship. *Füstchen* contained the electric power generator, fuel supply, powerful batteries, cylinders with various gases for breathing purposes, compressors and other equipment which ensured a continuous supply of air and power to *Heligoland*.

Heligoland's research programme, drawn up by the Biological Institute in conjunction with the Institute of Aeronautical Medicine,

Below left: Sponges are among the most ancient and primitive multi-cellular sea creatures. This bush-like specimen is only one example of their wide variety of colour and shape.

The long filaments of Physalia physalis, the Portuguese Man-o'-War, may be several metres long. They contain a vicious sting harmful to the swimmer who is unfortunate to come into contact with one.

was intended for a period of several years and included medical, physiological, biological, hydrological and geological research.

Heligoland spent the winter of 1969/70 under water, again demonstrating its reliability. Hurricane storms and sea currents did not appreciably damage either the undersea dwelling or its companion *Füstchen*. In the spring, after a ten-month stay on the sea floor, *Heligoland* was again visited by aquanauts who accompanied it back to the surface. *Heligoland* is still in service. One of the last undersea expeditions using the observatory took place in the Baltic. *Heligoland* was submerged for over three months, from April to July 1975.

The Makai Undersea Test Range

Now we travel from the Baltic Sea to the Makai Undersea Test Range,

domain of the Oceanographic Institute of Hawaii in the United States. According to the initial plans, an undersea city (aquapolis) was to be built here. It was to consist of several dome-like buildings, each 20 metres in diameter and about ten metres high, connected to one another and to the shore by a tunnel through which an electric mini-train was supposed to run. This elaborate project would have cost a vast sum of money if it had been realized, but it was replaced by a simpler and more practical enterprise, designed by the Swedish engineer, Gustav Falman, specially invited to Hawaii for this purpose.

According to his project, the undersea dwelling was to consist of three spacious pressure chambers, connected together and mounted on a catamaran-type pontoon. In submerging, the floats of the pontoon were to be filled with water.

Pilot whales can be found in all waters except in extreme polar regions. They migrate from cold to warm seas and back according to the season.

Together with special ballast compartments they held the dwelling firmly to the sea floor. Having enviable seaworthy properties compared to other undersea dwellings, this new home for aquanauts was also notable for the extent to which it was self-contained. It required neither a powerful floating crane nor a fleet of attendant vessels. The aid of a small boat was enough.

At the beginning of 1969 the

Despite international agreements to limit whale hunting, thousands of animals continue to be killed each year. The killer whale (left) preys on fish and other sea creatures, hunting in packs of 30–50.

undersea dwelling, built on the continent, was dismantled and delivered to Honolulu. There it was reassembled and towed to the pier of the Oceanographic Institute where, after festive ceremonies, Falman's creation was christened *Aegir* after the Norse god of the sea. Before being submerged to a great depth, the undersea dwelling was tested about 20 times in a shoal. During training, the aquanauts entered the emergency capsule which brought them to the surface. The "victims" of the simulated emergency went through decompression in a pressure chamber on shore.

Early in the morning of July 1, 1970, *Aegir's* aquanauts began compression after carefully closing all the hatches. The aquanauts received a quite complex assignment: to master, by practice, the operations required in servicing an undersea oil well. This entailed the manipulation of pipes weighing many hundreds of kilograms. The pipes and other material and equipment necessary for the experiment were accommodated on the deck.

A seven-armed Luida starfish raised on its slender arms in an egg-laying position.

By the end of the third day of underwater life all the planned work had been completed. The aquanauts entered the dwelling and, after getting ready for surfacing, started to blow out the ballast compartments. *Aegir's* bow quivered, detached itself from the bottom and reared steeply upward. But the stern was stuck fast in the floor. The aquanauts decided to repeat the process from the beginning but each time, instead of rising to the surface, *Aegir* reared up on end. All attempts to raise the dwelling failed.

On the morning of July 7, Falman gave the command to blow out the floats of the pontoon. This developed a buoyant force of 100 tonnes and immediately freed *Aegir* from

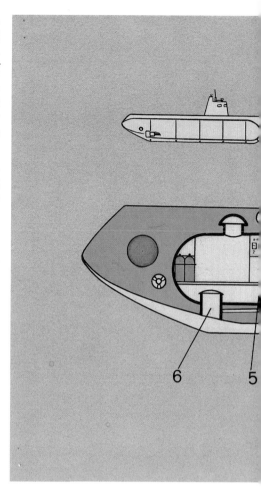

captivity. In white wreaths of foam, the undersea station broke the surface of the sea. It turned out that *Aegir's* unexpected imprisonment was due to the heavy oil well equipment. Owing to faulty arrangement of the three-tonne load on deck, the centre of gravity was displaced. Fortunately, the incident had no consequences and the undersea dwelling was in proper working order after passing this additional test with flying colours.

Japanese experiments

In recent years, Japanese scientists have also become undersea investigators. They began the initial stages of an extensive national project, called "Man in the Sea", to organ-

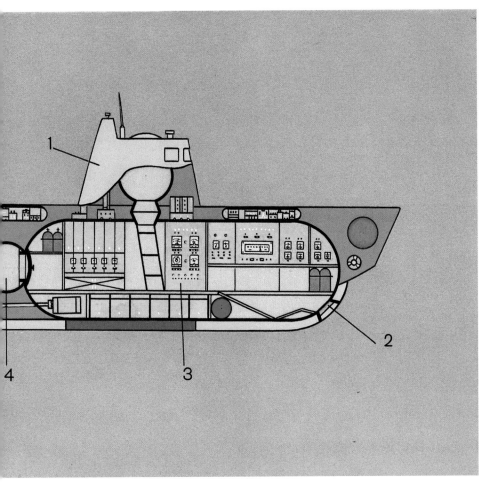

The French submarine Argironette designed by Jacques-Yves Cousteau.
1. conning tower;
2. porthole;
3. main compartment;
4. lock chamber;
5. diving compartment;
6. diving shaft.

ize a manned research station, *Sitopia*, on the floor of the Sea of Japan.

Japan displays especially keen interest in undersea investigations and has great expectations of developing the continental shelf. Japanese industry is based almost entirely on imported raw materials. The main aim of the "Man in the Sea" project was to try and change this situation, at least partly, by extracting commercial minerals—especially oil—from the sea floor.

The Japanese experts first set up their undersea dwelling at depths which did not exceed 100 metres. But in the near future they intend to approach the boundaries of the continental shelf and then cross them by establishing a large under-water dwelling at a depth of 250 metres. It is to house quite a large crew of undersea investigators who are to live on the sea floor for four or five weeks.

Meanwhile, other researchers of the sea depths envisage dwellings on the sea floor that can be moved from place to place, when the crew desires, without coming to the surface. There would be many obvious advantages to self-propelled undersea bases. Such dwellings must, of course, be completely self-contained and entirely independent of the shore and surface ships. This is exactly the type of underwater station that was developed by the designers of the Giprorybflot Institute of Leningrad and named *Bentos 300.*

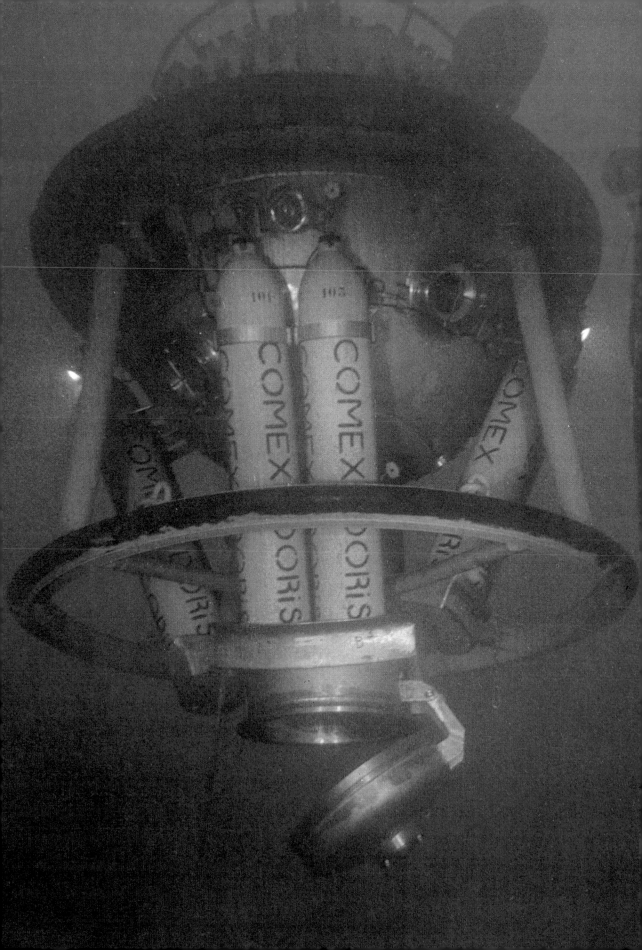

CHAPTER 5

AT HOME
IN
THE SEA

During the course of the expeditions of Edwin Link and Jacques-Yves Cousteau it became clear that thanks to the saturation effect there was another way to master the sea depths. It would be possible for the aquanauts to be based not on the sea bed but on the surface ship, or, to be more precise, in a pressure chamber stationed on the deck of the ship.

To increase working time under water it is not really necessary to keep a man on the sea bed, like a hermit. But it is essential to maintain the high pressure in which he lives. If this is achieved, then, in principle, it is not important where the aquanauts spend their leisure time or sleep, whether in an underwater dwelling or in a pressure chamber on the ship, or even in one on the shore. In all cases the saturation effect will be the same. The only difference is that people living in an underwater dwelling step out into the world of the sea depths immediately, whereas the aquanauts living on the surface have to go down to the sea bed and after work must rise to the surface again in a diving bell, with the necessary pressure in it maintained.

The Cachalot dives

Pressure chambers on board ship are not new. At least half a century ago they were in use. But the discovery of the saturation effect gave them a new lease of life. The compression complexes, which include a pressure chamber on deck plus a diving bell, have in recent times proved their worth. They have become reliable aids to man in his efforts to explore the continental shelf and the border areas of the continental slope.

Underwater dwellings are ideal for scientific research that takes time and requires constant attention. When it comes to assembly work and repairs on drilling and oil-extracting installations, or the laying and inspecting of oil pipelines and communications cables, it is more convenient to work with the help of diving bells and pressure chambers on board a vessel. What is more, compression complexes are cheaper, easier to make and simpler to service than underwater structures.

Significantly, one of the first complexes of this kind, called *Cachalot*, made its debut in a mountain reservoir, not in the sea. The task was to change a poorly designed installation at a hydro-power station there. While doing the job *Cachalot's* aquanauts had to spend a total of some 800 hours under water, each time returning to relax in a big 6·5 x 2-metre compression chamber on the dam.

The Comex diving bell. The shaft's hatch serves as an exit point for the aquanauts and is also used for docking with the pressure chambers on board the support ship.

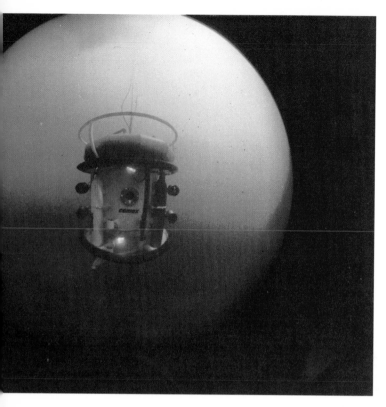

The next task entrusted to *Cachalot* was a more complicated one: to eliminate the consequences of an off-shore disaster. A storm in the Gulf of Mexico had toppled a drilling platform, which then sank to a depth of 60 metres. Salvage operations in the Gulf went on for several months and the crews of *Cachalot* changed 40 times.

Because of the growing demand for *Cachalot*, a second complex, *Cachalot 2*, was built. During the first two days of its service, this diving bell made ten descents to a depth of 183 metres, an impressive achievement. Formidable rivals to *Cachalot* soon appeared, underwater bells which could go down to even greater depths. One of these, called *Purissima*, was a two-stage bell built in the United States under Edwin Link's direction.

This high-pressure bell transports the aquanaut to the sea bed and back. It can also be used for the purposes of compression and decompression.

The new bell outwardly resembled the Soviet Union's *Sadko 2* station, consisting of two spherical compartments set one on top of the other, and both with portholes. The lower compartment is for the aquanauts, while the upper one, with a tightly sealed middle hatch, is used by the operator on duty. The air pressure in this compartment is normal. The operator instructs the aquanauts as they work, watching them through the portholes.

The next important development in the history of aquanautics and ocean-diving equipment was the appearance of the *ADS 4* complex. This provided work and rest conditions for the aquanauts over a two-week period, excluding the time required for decompression. The *ADS 4*, in its turn, was surpassed by a new complex, the *DDS Mk2*. This is a much bigger installation than any other hitherto designed, and more comfortable. Its four compartments can house 16 aquanauts simultaneously. Two deep-water

Down on the sea bed a diver measures the movement of drifting sediment.

diving bells look after their transportation down to the sea bed and back. Another of the most original complexes today is the *Dick Evans*, successfully used to assemble and repair underwater pipelines.

Divers on land

The engineers who developed deep-sea diving equipment have every reason to be proud of their achievement. The water barrier that blocked the way to the continental shelf has now been overcome. Great strides have also been taken by experts in underwater physiology. Their efforts to determine how the human body will react to underwater conditions have made it possible to use deep-sea diving equipment without impairing the health of the aquanauts. For example, preliminary experiments in land-based pressure chambers for training purposes enabled Soviet divers to conquer a "depth" of 300 metres as long ago as 1950.

The chronicle of deep-water descents simulated on land is full of vivid and, at times, dramatic events. In March 1968 a record "descent" was made by American aquanauts Carl Deckman and Daniel Fraser. They stayed at the simulated depth of 320 metres for five and 20 minutes respectively. But when the aquanauts came up, they learned that a similar achievement had been made by French colleagues. Moreover, the French researchers established a whole series of "descent" records on land. The pace was set by Henri Delauze, the head of Comex, a firm engaged in diving operations. Earlier, Delauze had worked with Cousteau and had the benefit of excellent training. Now he was acting on his own, hoping his experiments would interest the oil companies operating in different areas of the world's oceans. Delauze

The pressure chamber on board the Soviet research vessel Academician Orbeli.

Deepwater apparatus developed by Comex for servicing off-shore oil wells and underwater communications.

Lowering a diver wearing heavy diving gear into the water. Sometimes divers working on the sea bed prefer to use the traditional rigid suit.

began by building a compression complex in Marseilles.

In the spring of 1968 Comex began research with the *Physalie* programme, which received wide publicity the world over. The first experiment started on May 21. Delauze himself and his closest associate Rolff Brauer, a neurophysiologist, entered the hyperbaric compartment to establish personally whether man could survive at a depth of 335 metres, and to record scientifically their condition and their sensations during the experiment.

The target "depth" was achieved safely, and Delauze and Brauer decided to go even further, to 360 metres. The *Physalie 2* experiment began on June 11. This time the participants were Brauer and Roger Veurin, an engineer working with the firm. Again, the experiment was a success. The goal was reached, a record depth of 360 metres. Delauze and Brauer decided to continue their efforts and again increased the simulated depth.

At first the new experiment was unsuccessful. The sun had heated up the steel cylinders worn by the men,

and the temperature of the helium "cocktail" jumped to 50°C. Veurin was unable to stand the stuffiness and lost consciousness, although the "depth" was a mere 130 metres at the time. Brauer rushed to his companion's assistance and helped him to regain consciousness. But to continue the simulated descent was risky, so Brauer signalled for decompression.

On June 25 the experiment was repeated. The temperature in the compartment was no more than 36°C., and in an hour-and-a-quarter's time the researchers reached the "depth" of 305 metres. After that the descent proceeded at a slower pace until it exceeded 360 metres.

The results of Delauze's experiments persuaded the Elf oil company to reach an agreement with Comex, one of the first companies to do so. Delauze's firm contracted to help Elf extract oil in the stormy Bay of Biscay at depths of about 100 metres.

Meanwhile Delauze began another deep-water experiment in *Janus 2*. In Marseilles two teams of aquanauts on duty for alternate eight-day periods, carried out all the tasks assigned to them. The

Despite the fact that a storm is breaking out on the surface of the sea, work at the underwater laboratory takes its normal course.

(Opposite) The aquanaut's gas mixture is quickly consumed at great depths and it has to be pumped down through a pipe from the support ship or from a diving bell.

The graceful "flower" (top) spreading out its tender petals is in fact a sea anemone.

The sea anemone catches small fish and paralyses them with its poisonous tentacles.

A shoal of the many small fishes that settle in stony shallow waters.

The form and shape of the sea sponge vary greatly even among creatures of the same species.

physiologists made thorough checks of their state of health and behaviour, noting with satisfaction that the heliox, daily routine and the decompression regime had all been planned correctly. With everything going well, Delauze began the operation's decisive phase. On September 15, 1970, the ship *Astragale* dropped anchor in the Gulf of Ajaccio. Drilling equipment and instruments were lowered to a depth of 225 metres and the aquanauts went into action. By midnight on September 17 the pressure in the chamber had reached that of a depth of 200 metres and in the morning the first lowering began. The aquanauts inspected the working site and the equipment lowered beforehand, returned to relax, then took the diving bell down again to the sea bed.

The descents continued the next day, and for seven days after that. During the period each of the participants of the operation spent a total of 17 hours at a depth of 225 metres. At the time this was an unprecedented achievement and the *Janus 2* experiment came to a brilliant conclusion. Delauze had proved that his aquanauts could carry out an assignment at depths required by the geologists and oil experts.

A record "dive"

In England on March 3, 1970, two aquanauts John Bevan and Peter Sharphouse entered a compact steel pressure chamber and began a simulated descent. Within minutes the aquanauts reached a "depth" of 183 metres and stopped there for a day. Then they quickly simulated the next depth—305 metres. Again there was a pause. After a day's physical adjustment, the aquanauts continued their descent, this time

to 365 metres. There followed an hour's rest, after which they proceeded to 396 metres.

The physical state of Bevan and Sharphouse was excellent. After the third 24-hour stop to adjust themselves to the new conditions, the descent continued with the pressure in the chamber growing steadily. On the night of March 6 they reached a new record depth of 457·5 metres. Still the aquanauts felt well and were in a good mood. Of

explorers as Delauze and Brauer.

Underwater endurance

The exploits of Bevan and Sharphouse in England clearly indicated that it was too early to speak about a helium barrier. By that time the Comex firm had added three more test compartments to the existing ones, and in one of these a 1200-metre descent could be simulated.

An aquanaut watches a vibro-drill as it scoops up samples of sea bed sediment.

the ten hours spent at the final "depth" Bevan and Sharphouse slept for only two. The other eight were devoted to hard work. The two men carried out numerous and at times complicated tasks involved in making observations. They stayed five-and-a-half days at "depths" ranging from 305 to 457·5 metres. A mere twelve months before it would have seemed an impossible achievement to such bold

The aquanauts Bernard Roynet and Patrice Chemin were the leading figures in *Physalie 5*, a new experiment, which began on November 16, 1970. Pressure in the chamber occupied by the aquanauts had reached a simulated depth of 520 metres by the evening of November 19.

Intensifying the challenge to the sea depths, the veteran researcher Jacques-Yves Cousteau, by that time head of the firm SEMA, had put a shore-based compression complex into operation. In December 1971 the *Saturation 3* experiment was made in its compartments. Cousteau's aquanauts spent 24 hours at a "depth" of 400 metres the same length of time at 500 metres, and then another 24 hours at 400 metres.

On May 2, 1972, Henri Delauze began *Physalie 6*, an experiment featuring Patrice Chemin and Robert Gauret. This time a "depth" of 610 metres was reached. As previously, both the descent and the route back were carried out step by step. *Physalie 6* was completed on May 23 and its participants stepped out of their cramped metal quarters in excellent condition.

Methods of descending into the depths were being steadily perfected. In June 1974 the aquanauts Alain Jourde and Claude Bourdies, operating for Comex, stayed for 50 hours at the simulated depth of 610 metres. They also spent a much longer period—six days and nights—at a "depth" of 560 metres.

Deep-water descents

The past few years have seen many impressive experiments made by Soviet, American, British, West German, Italian, Swiss, Swedish, Dutch, Polish and Japanese specialists, operating both in special land-based compartments and in natural underwater conditions.

The physiologists, simulating deep-water descents in dry metal compartments and in ones filled with water, are testing many different kinds of "cocktails" for breathing. Recipes are varied both in terms of composition and in the proportions of the components. These changes are made depending on the depth to which the aquanaut is descending, whether he is coming

up or going down, and on the time he has stayed at a certain level.

In the quest for ideal breathing mixtures, hydrogen is frequently included. And so are exotic, inert gases, such as neon, krypton and xenon. Researchers feel that if these gases are used fully, or in part, to replace helium in air mixtures, this will in time enable man to descend to 1,000 metres and, possibly, even to 1,500 metres.

That man has still not realised his full potential in challenging the sea depths is shown by continuing experiments with animals. In several physiological laboratories there have been repeated experiments in which animals have been sent down to simulated depths of more than one kilometre. A staggering record was achieved, for example, by some pigs who endured a simulated 1,200-metres dive. But a team of monkeys outdid them, managing 1,500 metres. In both experiments the animals landed safely.

The complex of pressure chambers at the Soviet training centre of the Institute of Oceanography at Sky-Blue Bay in Gelenjik.

CHAPTER 6

EXPLORING

THE

DEPTHS

The role of trail-blazers in the sea depths where scuba divers cannot go, or can reach only at the greatest risk, is now played by submarines. Although research submarines are new arrivals on the scene, coming into use at about the same time as spaceships, they have already made a valuable contribution to science, studying ocean fauna, mineral resources, and the geophysics and geology of the sea bed.

The Severyanka

The Soviet government decided to dismantle one of the country's best naval submarines and then convert it into a superb science laboratory, an underwater branch of the National Research Centre for Fisheries and Oceanography. The centre had already acquired considerable experience of studying the sea with deep-water probes connected to vessels on the surface.

The submarine was called *Severyanka*, which means "the northern one", and the vessel's scientific activities were invariably carried out, as the name indicates, in the rigorous conditions of the northern sea. The *Severyanka* took part in many expeditions, covered tens of thousands of kilometres, and made hundreds of deep-water surveys.

One of the voyages took the scientists to the shores of Iceland and the Faroe Islands where Soviet trawlers were fishing at the time. During that voyage, which lasted many days, the researchers observed for the first time ever the herring's behaviour in wintertime. The fish were motionless and in most unusual positions: some were absolutely vertical, head up or head down; others were in a horizontal position, belly down, which could be regarded as natural for a fish; still others were lying belly up. But when the *Severyanka's* powerful lights were switched on, the sleepy mass came to life. The fish woke up and darted into the darkness.

Once the researchers noticed a huge shoal of herring, about a kilometre thick. But all attempts to come closer failed: the cautious fish would quickly move away as soon as the submarine's beacon lights spotted them. On another occasion the *Severyanka* dived under the deep-sea nets trawled by a surface vessel. From the submarine the designers of the fishing equipment could see how it worked, and even take film shots.

Thus, step by step, the depths of the Barents Sea and other northern waters revealed their secrets. The information obtained by the crew of the research submarine shed light on the way of life of cod and herring, their day-to-day and seasonal movements, and their reaction

Through the portholes of the West German submarine Mermaid I *(background) biologists observe fishes captured in their nets. Research into the behaviour of fishes has been of great benefit to the fishing industry and to designers of fishing equipment.*

to light and different kinds of noises. All this turned out to be of use to fishermen and the designers of new fishing equipment.

The *Severyanka* was soon to be joined by other science submarines, both in the Soviet Union and in other countries where new vessels appeared to probe the sea depths.

Tinro

A new research submarine, *Tinro*, whose name is derived from a Russian acronym standing for Pacific Centre for Fisheries and Oceanography, continued the work begun by *Severyanka*. *Tinro* studies the

On board the submarine, the hands of the depth gauge traced a sharply curved arc. The numerous instruments showed with precision the temperature, humidity and content of oxygen and carbon dioxide in the submarine's compartments. The acoustic rays pierced the mass of water. Sensors indicated the pressure outside, the water's temperature, transparency and salinity, and the tension of the earth's magnetic field.

On her expeditions, *Tinro* is accompanied by a base-ship, the *Ichthyander*. For, unlike the *Severyanka*, *Tinro* is a small submarine and during the intervals between its

Severyanka *at sea in the open expanses of the Arctic Ocean. The submarine has taken part in many expeditions and made hundreds of deepwater surveys.*

migration of fishes and other deepwater creatures. On board the submarine can be found not only sea biologists and ichthyologists, but hydrophysicists, acoustics experts and sea geologists.

After test voyages in the Black Sea the submarine set out for the ocean. The waves swept over its small superstructure and the vessel was rapidly submerged. Soon it disappeared altogether, and observers on the surface now turned to watch the screen of the hydrolocator to trace *Tinro's* movements.

underwater sailings it can be hauled on board the parent vessel. Though the submarine has only a limited amount of energy, oxygen and food supplies for the crew, it can do what the much bigger *Severyanka* could not—go down 400 metres into the ocean depths.

The Denise

Two men in divers' gear push their way through the narrow opening of the submarine *Denise* and tightly close the hatch behind them.

Their hands operate the buttons and keys of the control panel, switching on the regeneration apparatus, and testing the cameras, cine-cameras and tape-recorder. Then the *Denise* is picked up by a crane, lowered gently into the water and released. The submarine slips down into the sea. The fathometer is switched on to send out and receive signals which give warning of possible obstacles ahead.

Mention has already been made in this book of the *Denise*, veteran of many of Jacques-Yves Cousteau's expeditions. Oval, squat, with bulging, illuminated portholes and mechanical claws, the vessel looks like a living creature straight out of a science-fiction novel. But the exotic appearance of the ship, designed by engineers Mollarde and

The Soviet submarine Osa 3 submerging in the Baltic Sea. It is designed to be highly manoeuvrable.

whereas the latter can descend to 300 metres, these one-seater midget submarines will, as their name indicates, go down to 500 metres.

Osa 3

At a settlement near Vyborg, on the Gulf of Finland, the maiden voyage took place of another submarine, the *Osa 3*, designed by engineers of the Moscow institute who also produce vessels for fishing fleets. These engineers, turned hydronauts, themselves tested the original submarine. *Osa* is an acronym derived from the Russian words meaning Inhabited Stabilized Apparatus. The figure three stands for the number of places on board. *Osa 3* is not only one of the latest, but is also one of the most interesting research submarines, even more versatile than *Denise*. Despite the *Osa 3's* formidable displacement of 12 tonnes, it can move easily in all directions, and is able to remain stationary despite the ocean currents. Work on *Osa 3* earned the Moscow designers some 24 patents. But the regeneration system is almost entirely borrowed from space engineers.

Osa 3 operates under the Soviet National Research Centre for Fisheries and Oceanography. Therefore

Laban, together with Cousteau himself, is dictated by strictly practical considerations. The combination of a turtle-like hull and a special engine with revolving nozzles shooting out water jets, guarantees the *Denise* complete freedom of action under water. The vessel can with equal ease move forwards or backwards, to the right or left, revolve round its axis, remain motionless, and achieve the necessary inclination to climb or dive. Since 1960 the *Denise* has made some 2,000 descents and no other submarine can equal her achievements.

Later, Cousteau's fleet received two new additions, the twin submarines SP 500. Their design is based on that of the *Denise*, but

The crimson multi-armed starfish is in no danger from other creatures as its body is poisonous to them.

its main task is to carry out biological research for the fishing industry. But the submarine is well able to perform other tasks. Its electro-hydraulic manipulator, ably handled by the hydronauts, does miracles in assembly under water, and meticulously chooses mineral samples. *Osa 3* can penetrate the domain of the continental shelf and its slopes down to 600 metres.

Stars of the depths

The appearance of the *Denise* and her sister ships, the SP 500 twins, attracted worldwide interest. Oceanographers everywhere welcomed them as convenient multi-purpose vessels. They are regarded as the progenitors of a whole family of submarines that has appeared in recent years in France and elsewhere. The United States *Deep Star* submarines, indexed 2,000, 4,000, 12,000, 13,000 and 20,000 to denote how many feet they can descend, are good examples. Jacques-Yves Cousteau contributed to their design and gave useful advice to his American colleagues. The first of these submarines, the *Deep Star 4,000*, was tested at Marseilles, before being airlifted to the United States.

Many interesting observations

were made during descents by *Deep Star 4,000*. One of these was an unexpected encounter with a huge, hitherto unknown, species of fish, some 10-12 metres long, with a huge grey-dotted body and eyes as big as dinner plates. The fish was calmly posing in the light beamed from the submarine and could be clearly seen through the portholes. So unusual was the sight that when the crew returned to the surface, they did not immediately relate what they had seen at a depth of 1,250 metres in the San Diego depression. Only

Mermaid 3, a West German research submarine with a pressure chamber which can accommodate three aquanauts.

The birth of a tiger shark. It is among the most voracious and aggressive underwater creatures.

Between the passenger gondola and the glass-fibre fairing, which lets through water, there are batteries, electric engines, water and mercury tanks. The free space is taken up by the foam coming out of countless numbers of tiny, hollow glass spheres, almost twice as light as sea water. The arrangement has proved its worth and has functioned brilliantly even when the pressure amounted to 600 atmospheres.

Adventures of the Alvin

The *Alvin*, a veteran American submarine, was intended for biological, geological and physical research. It could descend to 2,000 metres. In 1966 the vessel performed a very responsible mission. Two United States Air Force planes had collided in mid-air at Palomares in Spain and, shattered, had plunged down to earth. One of the planes had four unarmed nuclear bombs on board. Three of these fell to earth and were recovered. The fourth bomb landed in the sea. Of the search vessels sent to the area it was the *Alvin* which found and recovered the bomb, locating it after 34 dives.

Returning home, *Alvin* resumed its research work. The following year the submarine made 200 descents: it took photographs of iron and manganese fields over a kilometre down in the Atlantic, measured the speed of the Gulf Stream current at different levels, and studied the behaviour of the sound-dispersing layer at Cape Hatteras.

In the fifth year of its activities disaster befell the *Alvin*. When the submarine was being lowered into the water, the hoist cables broke and, with hatch open but fortunately with no one on board, *Alvin* plunged into the ocean depths. For nearly a year the submarine was lying on the ocean bed, about a kilometre down,

The French submarine Ciena. *In 1974 this vessel together with the* Alvin *and the bathyscaphe* Archimède *made a series of dives to survey the mid-Atlantic underwater mountain ranges.*

after another giant fish of the same appearance was photographed deep under the water, did scientists conclude that this was a Greenland shark.

Unlike the somewhat squat *Denise*, *Deep Star's* body consists of two welded steel semispheres with a glass-fibre fuselage fairing, with beacons and mechanical claws on the sides. Instead of jet engines the vessel has ordinary screw propellers.

becoming a home for fish and other sea creatures. Months later, the submarine was refloated and resumed its research work in the ocean depths. Recently, another similar vessel, the *Alvin 2*, was built. This remarkable research submarine has already made several hundred deep-water descents.

Canadian submarines in the *Pisces* series are somewhat more sophisticated than the *Alvins*. A *Pisces* consists of two sturdy spherical bodies. The bigger one houses the crew, while the smaller one is for equipment. *Pisces 1* can descend to 975 metres. The vessel is equipped with a manipulator and boring equipment and, as with all the other vessels in the series, can do the job of oceanographer, underwater prospector, drilling expert and assemblyman, as well as check and repair underwater cables.

An angelfish (top left) among acropora corals. Ascidians, or sea squirts (bottom left) lead a solitary life, clinging to rocks and cliffs.

The Soviet submarine Argus.
1 and 6. main ballast tanks; 2. hull fairing; 3. hatch; 4. signal light; 5. pilot's seat; 7. observer's seat; 8. basket for collecting samples; 9. mechanical hand.

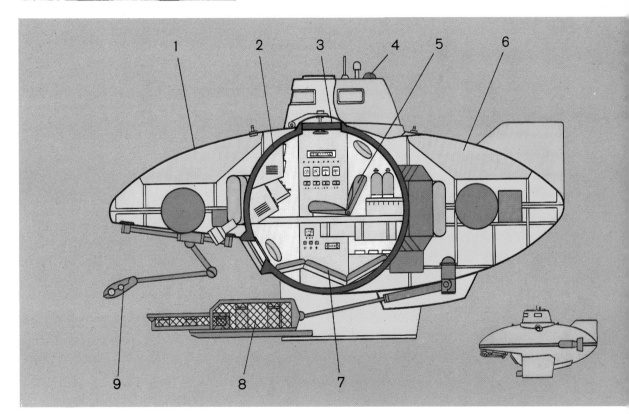

Pisces 2 and *Pisces 3* can go down to 1,000 metres. More recent submarines in the same class can descend to 2,000 metres. Some of the best research submarines of this kind, the *Pisces 4* and *Pisces 7*, were acquired by the USSR Academy of Sciences and are now operating for the southern branch of the Institute of Oceanography. Soviet researchers are well pleased with the submarines: they are sturdy, reliable and easy to handle.

Under the sign of Perseus

The scene is the cold, inhospitable Barents Sea. Somewhere far away beyond the horizon lies the Kola Peninsula. High above, the flashes of Aurora Borealis pierce the greyish skies and reflect themselves in the waves on which a vessel is swaying. The people on board are peering anxiously to sea, trying to discern something in the darkness. Then someone notices a light spot in the water. With every passing second the spot grows more distinct. Presently a submarine appears on the surface, bearing a blue triangular flag with seven white stars. This represents the constellation of Perseus, the emblem of the Polar research centre for fisheries and oceanography. It is under the command of this centre, stationed in the Arctic city of Murmansk, that this new deep-water research vessel, *Sever 2*, operates. Sever in Russian means "the north".

The American submarine Alvin.
1. hydraulic drive for propeller;
2. manoeuvring propeller;
3 and 5. deepwater lights;
4. television camera;
6. batteries.

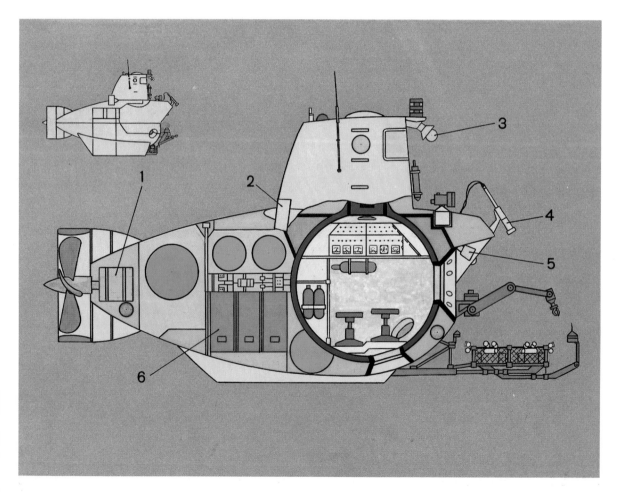

The *Sever 2* is a large underwater vessel, 12 metres long and three metres wide. The commander, engineer and the two researchers on board are all provided with convenient, reclining seats. There are seven portholes, exterior searchlights and a battery of cameras and cine-cameras for probing the underwater world. *Sever 2* is packed with every kind of equipment. The submarine can rest on the ocean bed, or be moored like a dirigible, with the guide rope serving as an anchoring device. The vessel's mechanical manipulator can meanwhile scoop up samples of soil and deep-water flora and fauna. *Sever 2* can also drift with the sea currents, or sail in any direction at any depth down to 2,000 metres.

Sever 2 carries out an extensive range of scientific research in northern seas, where it now operates. But the vessel's main job is to study the laws of fish life and development, discover fish shoals, and search for new species of fish and sea animals that could serve as food. Trawling that *Sever 2* has done in depths of one kilometre confirms the presence there of great quantities of valuable species of fish, and fishing trawlers are already catching them.

The *Alvin* afloat. It is named after the American oceanographer Allen Vine.

Sever 2 *at sea. The submarine's conning tower bears the Soviet State emblem.*

The abyssal zone

In recent years new vessels have been launched to probe the depths of the continental slope, vessels such as *Deep Jeep*, *Moray*, *Dowb* and *Deep Quest*. Each has many points of interest and is a work of skilled design and engineering. The vessels incorporate many of the latest advances in aerohydrodynamics, electronics, metallurgy, metal-processing, electric engineering, bionics, and even aviation and cosmonautics.

Deep Jeep, which looks like a giant jellyfish, is made of the finest steel, glass and plastic. It performs the widest range of tasks and can, in turn, act as a biological prober, a prospector, an inspector of underwater oil pipelines and rig structure, or even an archaeologist as it searches for buried objects in the sea

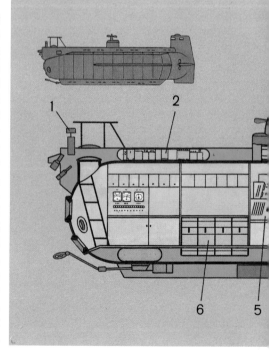

bed. *Deep Jeep* can reach a depth of 2,000 metres. The torpedo-like *Moray* can descend even further, to 3,000 metres. *Moray* has two spherical compartments, one for the crew and the other for equipment.

In appearance *Deep Quest* resembles the fuselage of a wingless fighter plane. Immediately after launching, the submarine, with a crew of two researchers and two engineers, descended 2,485 metres to reach the bottom of the Valero hollow in the Pacific. The trip to the continental slope lasted four and a half hours. The descent began at 7.00 a.m. local time, and at 11.37 a.m. the submarine carefully settled down on the ocean bed, raising a cloud of silt.

But the depth reached that day is not the limit for a submarine. The ocean's abyssal zone is more than 2,000 metres down. Yet it was reached many times by the submarine known as the *Aluminaut*. Undoubtedly one of the world's

best underwater craft, it can only be rivalled by the *Denise* and the *Alvin*.

The *Aluminaut* has rendered valuable service to scientific research. This applied particularly to sea geology, the study of mineral resources which lie on the ocean bed. Thus, when probing the Blake Plateau off northern Florida, the oceanographers on board the submarine discovered hitherto unknown deposits of ferro-manganese concretions. These deposits proved to be truly enormous—huge, black, round-shaped stones seemed to pave the ocean bottom. It was like a street swept clean by the Gulf Stream, so clean that the *Aluminaut* was able to move along it on its rubber wheels to save energy.

On another occasion, during its continuous probes in the Atlantic,

Soviet underwater craft Gvydon. *Its shape is unique and resembles the tube of an artillery shell.*

The Aluminaut, *an American submarine capable of reaching the abyssal depths.*
1. *radar;*
2. *research equipment;*
3. *oxygen cylinders;*
4. *emergency ballast;*
5. *central control station;*
6. *batteries.*

One of the Canadian Pisces *submarines* now flying the Soviet flag. In 1977 it was used to investigate the bed of Lake Baikal, the world's deepest natural lake.

Aluminaut's crew discovered the anti-Gulf Stream, a current moving from north to south, not the other way round, as one would expect.

In the depths of the Gulf Stream

An unusual trip involving a 2,400-kilometre drift with the Gulf Stream was made by the research submarine *Ben Franklin*, designed by the Swiss engineer Jacques Piccard. He first thought of such an expedition when the *Auguste Piccard*, a multi-seated excursion submarine which made hundreds of dives in Lake Geneva, was launched.

It was with that spacious vessel that Jacques Piccard originally decided to put his plan into practice, after the submarine had been converted into an oceanological observatory. But then he abandoned the idea and instead built a new

vessel, one that would be more suitable for lengthy underwater voyages. The *Ben Franklin* was the result, a research submarine that had a seven-day supply of electricity, a gas mixture for breathing, and a supply of fresh water and food for the crew.

For underwater sailing the submarine was equipped with four powerful propeller screws. But these were turned on only when it was necessary to alter the depth of descent, or to return to the charted route if the vessel had, for one reason or another, steered off it. The *Ben Franklin* could remain at a fixed point, or settle down on the sea bed.

The drifting underwater observatory was manned by six persons, including the expedition's head, 45-year-old Jacques Piccard. When sailing day came, at a point situated off the Florida coast, all hatches on the vessel were sealed tight and it slowly descended into the depths of the Atlantic to begin its underwater drift.

Although the Gulf Stream is called a warm current, the hydronauts could scarcely regard it as such. At a great depth the water's temperature went down dramatically, as did the temperature in the submarine's living compartments. The hydronauts had to put on heavy clothing to keep warm and electricity had to be conserved.

In the very first days of the underwater drift, when heading for the area of Cape Canaveral, the researchers noticed through the portholes a giant jellyfish, some ten metres long. But barely had they passed this creature when they saw two big swordfish. One was particularly ill-tempered and aggressive. It suddenly swerved round to attack the submarine. The vessel shuddered but fortunately no damage was caused. The submarine's steel exterior proved strong and reliable. Later on the hydronauts saw that a big octopus had fastened itself to one of the portholes. It stayed there for several hours, taking a free ride.

At times the researchers, turning the outside lights on, would discover that a host of big sea creatures was accompanying their slow-moving vessel. Sometimes shoals of tunny fish would be swimming along, sometimes dolphins, and sometimes even whales. On the twentieth day of the expedition the *Ben Franklin* found itself surrounded by blue and grey sharks, and hammerhead sharks. These were all gazing at the submarine, which must have seemed a weird creature to them, but none took any hostile action.

On the thirtieth day of the drift the *Ben Franklin* was still heading north, but the underwater journey was already coming to an end. Some 480 kilometres off the Florida coast the *Ben Franklin* switched on all four propeller screws and slowly surfaced. Behind lay 731 hours, 28 minutes of voyaging in the Gulf Stream's depths, with some three million different measurements made automatically by the instruments of the drifting observatory.

Exploration by submarine

In the Soviet Union, some of the most advanced research submarines also contain hyperbaric, or high pressure compartments, and a voyage on such a craft is described by one of the scientists on board. "There was a rough, wintry sea when our unusual submarine with its three rooms for the underwater explorers set out. Through the portholes of the pressure chambers it was easy to see how much at home the aquanauts felt. They were seasoned explorers of the sea depths, having already spent many days under water. For a month the aquanauts worked on the sea bed. For the first time, man was leaving a submarine at great depths to remain for periods of several hours in the open sea. This expedition marks a new stage in the development of underwater exploration."

The main advantage of such deep-sea dwellings is the very fact that they are part of large submarines. This opens up interesting scientific possibilities and there are many obvious benefits for the aquanauts themselves. Everything needed for life under water is supplied continuously to spacious and comfortable apartments. The aquanauts receive heliox, heat, light, hot water and food.

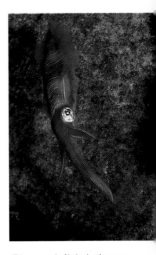

The cuttlefish belongs to the genus Sepia. It has an inky secretion from which is prepared a brown pigment used in water-colour painting.

CHAPTER 7

BATHYSCAPHES—
SHIPS OF THE
OCEAN DEPTHS

Bathyscaphes, ships capable of reaching the bottom of the ocean even in its deepest depressions, trace their origins to the earlier bathyspheres, which were attached by cable to surface ships. This probably accounts for the fact that bathyscaphes, or "super-depth ships" preceded by a number of years the appearance of submarines like *Denise, Alvin* and *Deep Quest.* Moreover, much in the structure and operation of these submarines was clearly borrowed from the bathyscaphes.

FNRS 2 and 3

In the 1930s, years marked by Barton and Beebe's triumphant submersions into the depths of the Atlantic on an attached apparatus, Konstantin Tsiolkovsky, the founder of scientific cosmonautics, put forward an interesting idea in regard to the conquest of oceanic depths. The Russian scientist suggested the building of a bathysphere that would have a certain reserve of buoyancy. In submersion this buoyancy would be balanced by ballast. When the descent had been completed, the extra weight would automatically be discarded and the bathysphere would rise to the surface.

Tsiolkovsky's idea was further developed by the Soviet shipbuilder Yu. A. Shimansky. Shortly before World War II he designed a self-propelled apparatus capable of going down to a depth of several kilometres. Unfortunately, the war prevented the realization of this interesting project.

In the post-war years great achievements in conquering super-depths were scored by the Swiss scientist Auguste Piccard. The bottom of the deepest depression in the world's oceans was reached in a bathyscaphe of his construction.

By that time Piccard had already earned world fame as an explorer of the stratosphere. He had twice risen to the record altitudes of 15,780 and 16,300 metres to study the mysterious cosmic rays which presented such a puzzle to scientists in the 1920s and 1930s. He made these trips in the world's first high-altitude aerostat FNRS 1 with a hermetic gondola of his own design.

It was from aeronautics that Piccard borrowed the principle of the bathyscaphe, which is often referred to as an underwater dirigible. Nor is this merely a metaphor. The dirigible is raised aloft by the light gases, helium or hydrogen, which fill its casing. In a bathyscaphe this function is performed by petrol, which is considerably lighter than sea water. A bathyscaphe with a steel gondola weighing several tonnes and capable of

The bathyscaphe Archimède *designed by Georges Houot and Pierre Willm has made many descents to depths of over 9,000 metres.*

The bathyscaphe Trieste *is remembered principally for its exploration of the Mariana Trench.*

Archimède *returning to base after a record descent off the shores of Japan.*

withstanding the pressure of the water at a depth of several kilometres beneath the sea surface goes down under its own weight, and floats up, like an aerostat, as soon as the ballast is discarded.

Piccard called his bathyscaphe FNRS 2, just as he called his aerostat in earlier years, after the Belgian establishment *Fonds National de la Recherche Scientifique Belge* which financed both ventures.

After preliminary tests in shallow waters, the FNRS 2 made a descent to a depth of 1,400 metres in the Atlantic, in the vicinity of Cape Verde Islands. However, after conquering this great depth, the bathyscaphe exhibited unexpected weaknesses in the face of an ordinary storm, which gathered on the surface and did it considerable damage during the towing. In fact, the damage was so great that it was not worth trying to repair the ship and it was scrapped. Its gondola, however, remained intact and was inherited by the next model FNRS 3, built jointly by Belgian and French engineers with the participation of Piccard.

On February 13, 1954 the FNRS 3 bathyscaphe piloted by the Frenchman Georges Houot and Pierre Willm reached the record depth of 4,050 metres. From time to time sharks swam into the explorers' field of vision. At first the two refused to believe their eyes. At that time nobody imagined that these big predators could live at such enormous depths. The bathyscaphe enabled man to peer into the unknown world where formerly people had no access.

In the Mariana Trench

One morning in January 1960, the ship *Wondenks* left the harbour of the Island of Guam in the early

hours, leading in tow an unusual craft. It was the bathyscaphe *Trieste*, whose crew consisted of the engineer Jacques Piccard, the son of Auguste Piccard, and Lt. Donald Walsh of the United States Navy. They were going to explore the Mariana Trench. The day was dull and drizzly and the *Trieste* on its long cable was tossed about by the big, menacing waves. At 8.23 a.m. the long-awaited submersion began.

Five hours after the beginning of the descent the *Trieste* landed on the hard, even bottom. The beam of the searchlight revealed a fish. Flat like

Jacques Piccard, whose dives in the Trieste *have taken him to depths of 11,000 metres.*

Starfish underwater dwelling (France)

Sadko 3 underwater dwelling (USSR)

Chernomor 2 underwater laboratory (USSR)

MAI 3 submarine (USSR)

Sea diver with aqualung of compressed air

Sea Lab 2 underwater dwelling (USA)

Shallow-water submarine

Auguste Piccard mesoscaphe

Sea diver in flexible diving suit

Précontinent 3 underwater dwelling (France)

Severyanka submarine (USSR)

Sea diver in rigid diving suit

Kuroshio 2 submarine (Japan)

Denise diving saucer (France)

Deep Diver underwater apparatus (USA)

Tinro 2 underwater apparatus (USSR)

Ben Franklin submarine

Nuclear-powered submarine

Argus underwater apparatus (USSR)

Star 3 underwater apparatus (USA)

Sever 1 hydrostat (USSR)

Bathysphere

Deep Star 4000 underwater apparatus (USA)

Deep View underwater apparatus (USA)

Alvin underwater apparatus (USA)

Sever 2 underwater apparatus (USSR)

Deep Quest underwater apparatus (USA)

Remote-control *Crab* underwater apparatus (USSR)

Aluminaut submarine (USA)

Buoy station

Archimède bathyscaphe (France)

Trieste bathyscaphe

Deepwater cine-photo robot

Automatic bathyscaphe

Red soldierfish guarding its den. It will attack other creatures with its spread-out sharp fins.

Lucernaria are jellyfish which do not move about but attach themselves to algae. The colour of these sea animals always corresponds to the colour of the algae with which they live.

The anemone, or clownfish, belonging to the genus Amphiprion, lives in association with certain sea anemones. The fish is immune from the poisonous tentacles of the anemone and shares its food.

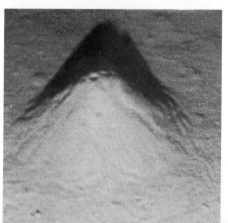

Some creatures living on the sea bed scurry for safety by burrowing beneath the sand. This picture was taken from an underwater vessel at a depth of over 1,000 metres.

a plaice, it was lying on the ground and watching the bathyscaphe with round, phosphorescent eyes. Then it swam away unhurriedly and disappeared into the darkness. The hydronauts discovered another live creature, a fragile-looking shrimp. This put an end to arguments as to whether there was any life at great depths of the ocean.

The Archimède

Several years after the building of FNRS 3, Georges Houot and Pierre Willm designed and tested another bathyscaphe, which they named *Archimède*. Like the *Trieste*, *Archimède* can submerge to any depth. It was the first to reach the bottom of the Puerto-Rican Trench, the 3,200-metre deep chasm in the Atlantic. In the Pacific, near the shores of Japan, the *Archimède* made several submersions to depths of over 9,000 metres.

The Soviet geophysicist Valeria Troitskaya was on board the *Archimède* during one of its submersions in the Mediterranean. She was the first woman to have travelled on a super-depth ship. Valeria Troitskaya had come to France in connection with her studies on the earth's magnetic field. These studies used to be carried out on dry land and in the atmosphere. Now she was to conduct experiments on the sea bottom. "The idea that a fully autonomous vessel, isolated from the sea surface and the ground, would carry us down to the depth of several thousand metres thrilled me extraordinarily," she said.

"The descent to a depth of 2,500 metres took up two hours. We spent five hours on the sea bottom, moving just above its surface at a speed of four or five knots. We saw fish of most fantastic shapes and colours. Some had fins which resem-

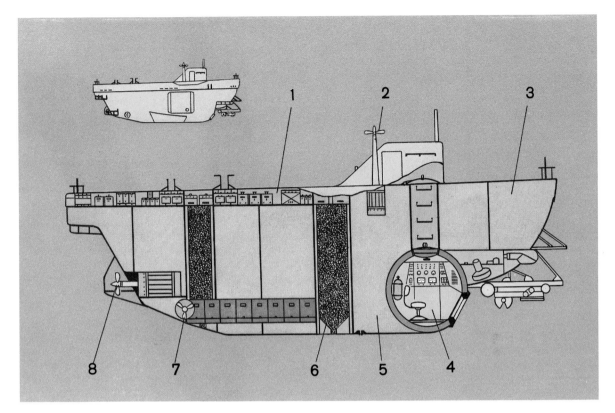

bled legs, and they stood on them like lizards, raising their heads to watch our bathyscaphe. Here, at these depths, we also saw objects that reminded us of mankind—bottles, bits of metal and dozens of shiny coins tossed into the sea by tourists 'for luck'."

Some time ago the *Archimède* joined two submarines, the American *Alvin* and the French *Cyana* to conduct a geological and geophysical survey at depths of some 3,000 metres of a large sector of the rift valley to the south-east of the Azores. The expedition lasted several months. Its most sensational result was a brilliant confirmation of the hypothesis concerning lithospheric plate tectonics, set forth in the first chapter of this book.

During dozens of submersions to the sea bottom, protected by the steel plate of the submarines and the bathyscaphe, the scientists were able to observe molten bazaltic lava

pouring out through fissures in the oceanic bottom, causing the water to boil. As the lava cooled, it often formed hollow spheres and pipes, which were immediately crushed by the pressure of the water. The researchers watched the glass-like surface of the lava change its consistency and saw hot salty springs gush forth from the bottom.

The mechanical hands of the submarines brought up to the surface several hundred kilograms of fragments of solidified lava. They also took dozens of samples of thermal saline solutions which oozed from the earth's interior. Thus it was confirmed once again that rifts at large depths in the ocean are the most dynamic and the most pliable areas of the earth's crust. Here young crust is born, here the molten substances of the upper mantle emerge to the surface. It is here, in rift zones, that we can feel the pulse of our planet.

The bathyscaphe Archimède.
1. scientific equipment;
2, 7 and 8. screw propellers;
3. float hull;
4. central gondola;
5. gas tank for manoeuvring;
6. bunker with emergency dry ballast.

CHAPTER 8

CONSERVATION
OF THE
OCEANS

Today, the interdependences in the biosphere, the processes that occur in the environment as the result of man's activities and their final effects are the focus of attention of scientists, statesmen and public figures the world over. Only a thorough study of the processes at work and a strict observation of the scientists' recommendations may help us to avoid the dangerous consequences of man's influence on nature.

The lesson of Torrey Canyon

We find ourselves in a paradoxical and largely unenviable position: while formerly nature was often inimical to man, now man is often inimical to nature. The dramatic events which took place near the shores of England and France in the spring of 1967 are still fresh in people's memory. The huge tanker *Torrey Canyon* ran aground on the Seven Rocks reefs in the North Sea with a cargo of 120,000 tonnes of raw oil. Battered by the waves of a stormy sea the tanker broke apart and the oil began to pour out into the water. To stop the pollution of the sea, it was decided to set fire to the tanker. Fighter and bomber planes from Great Britain dropped hundreds of explosive and napalm bombs and rockets on the vessel. A huge conflagration rose over the

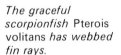

Opposite: Coral reefs in tropical waters provide the underwater photographer with a fascinating array of colourful subjects.

The graceful scorpionfish Pterois volitans *has webbed fin rays.*

This butterflyfish and a grouper live in crystal-clear coral waters and are noted for their highly variegated colours.

The spikes of Acropora *corals can rip the bottom of a small vessel.*

completely in the polluted zone, which stretched for hundreds of kilometres in width and was 10-15 metres in depth. On the coast of Cornwall alone the black breakers tossed out about 40,000 oil-smeared, dead sea birds.

The disaster in this populated corner of the world's oceans caused concern to millions of people, especially those who depend on the sea for their living.

Pollution in the sea

Unfortunately, accidents like the one which befell the *Torrey Canyon* are not isolated instances. Similar disasters have occurred scores of times. But, curiously enough, far more damage is done to the sea by tankers which reach their destinations safely. After unloading, these vessels wash out their tanks with sea water, discharging a lot of oil products into the sea. Not so long ago this was done openly in port. Today this is strictly prohibited. But in the open sea you still come

The coat of the fur seal has a thick under-fur which is very valuable. Once hunted almost to extinction, fur seals are now protected by international laws.

Sea worms prey on small crustaceans, molluscs and other worms.

The tail thorns of the surgeonfish are so sharp they can be compared to a scalpel.

Atlantic, which was enveloped by clouds of black, acrid smoke.

But time had been lost. When the bombing was begun, 60,000 tonnes of oil were already in the sea. The jelly-like brown mass, oil mixed with sea water, was carried by the current to the British coast. A fortnight later the oil reached the shallows of Brittany in France. The owners of seaside boarding houses and hotels on both sides of the English Channel were threatened with ruin. Thousands of fishermen were faced with the prospect of losing their livelihood.

Studies some time after the *Torrey Canyon* disaster confirmed the worst fears. Crabs, lobsters, molluscs, fish and echinoderms perished almost

across huge oil lakes. This means a tanker has been cleansing itself at the expense of the sea and its creatures.

According to the assessment of specialists, nearly a million tonnes of oil are poured out annually into the world's oceans by tankers and other ships.

Particular danger is also entailed by the discharge of chemical waste, because of its great toxicity and stability. The waters of the world's oceans are in fact a single sea, and phenomena and processes occurring in one part of it are sure to affect other parts sooner or later. Evidence of this are traces of DDT found in the bodies of Antarctic penguins. Already it has become necessary to forbid fishing along huge sections of the coastal shelf of a number of economically developed countries because the sea foods obtained there may be poisoned. Many rivers, in fact, have become huge sewer arteries carrying all sorts of industrial waste into the sea.

Radioactive isotopes present a danger that is all the more deadly for being concealed. They have no colour, smell or taste. Once in the sea, they are absorbed by plankton and seaweeds and thus infect the initial links of the ocean's great food chain. We know quite a number of species of fish, molluscs and crustaceans which accumulate radioactive isotopes. There is no defence against radioactive substances which have made their way into the sea. All that remains is to wait until they disintegrate of themselves. And sometimes it is a very long wait: the half-life of many dangerous radioactive substances lasts dozens, hundreds and even millions of years.

According to an International Convention precluding the pollution of sea waters, concluded in

1954, all water containing oil products and their waste should be poured into the holds of special "sanitary" ships or pumped into coastal sumps. The Convention declared the North, Baltic, Norwegian and some other seas forbidden for the discharge of polluted waters. Later, at a conference in London another, even stricter convention on the protection of waters, especially in the closed-in Baltic Sea, was adopted. Measures have also been taken to stop the pollution of the Mediterranean Sea, whose waters, as with the Baltic, take a long time to renew due to the narrowness of the straits.

Great attention to the protection of sea waters is paid in the Soviet

The sea elephant is now a rare animal. Intense hunting nearly exterminated the species.

That dolphins were friendly to man and easily trained has been known since ancient times.

Ascidians or sea squirts in the Sea of Japan.

Union where very rigid laws to this effect are being enforced. All ships of river and sea fleets of the country are supplied with tanks for the collection of polluted waters. Not a gram of water used for the washing of the tankers registered in Soviet ports is discharged into the sea.

In Odessa, for instance, a "sanitary" ship has been in operation for many years now. It services the ships which arrive at this port from all parts of the world. In the first four years of its service alone 10,000 tonnes of oil were recovered. When the holds of the whaling flotilla *Slava* were cleaned with its help, another 1,240 tonnes of whale fat were recovered after unloading.

A very important event took place on August 5, 1963, when an international treaty on the banning of nuclear tests in the atmosphere, cosmic space and under water was concluded and subsequently signed by nearly all countries in the world. This was a great achievement of good sense and good will on the part of nations and governments. Several years later, on February 11, 1971, peace-loving humanity scored another victory—the agreement to ban the storage on the bottom of the sea and oceans of nuclear and other mass destruction weapons.

International co-operation

The offensive against the ocean is advancing along a wide front. Serious dangers threaten the sea fauna not only through environmental pollution but also through excessive fishing and hunting for sea animals in the traditional areas. Here, too, international co-operation can be effective. In 1946 the Washington Whaling Convention was concluded, which regulated whaling in the Antarctic. Scientists are now urging that a limit should be set on fish catches for each fishing area and that fishing should be prohibited during the spawning period.

In 1956 the USSR and Japan agreed mutual obligations to protect and increase the reserves of salmon, crabs and herring in the north-western part of the Pacific. In 1957, the USSR, Canada, Japan and the United States agreed on the the preservation of fur seals in the northern part of the Pacific. An agreement between the USSR and the United States about joint catching of the most valuable kinds of crab in the eastern part of the Bering Sea has been in force since 1966. In September 1973 representatives of seven countries, the USSR, the German Democratic Republic, Poland, Sweden, Denmark, the German Federal Republic and Finland, signed an agreement on the protection of fauna in the Baltic. Many similar agreements could be named and all of them proving effective.

An important factor in the preservation and increase of oceanic fauna is the search for unknown areas of habitation of fish and sea animals and the expansion of fishing at great depths; as well as the artificial breeding of fish, molluscs and crustaceans, as described in Chapter One of this book.

It is gradually being brought home to man that if he wants to continue to exploit the ocean's fish reserves he must pass from primitive hunting to civilized husbandry. It is necessary to feed fish, to add nitrogen and phosphorous salts to some sectors of the world's oceans to help the reproduction of valuable species of fish and to destroy the predators. After all, agriculture is based on correct cultivation of the soil, the use of fertilizers, care of the fields, or, in animal husbandry, care of the stock. Similar activities are becoming necessary in the ocean, even though they involve greater difficulties due to the vastness of its expanses and the incessant shifting of the waters.

The overall conclusion that can be drawn from everything that has been said in this book is as follows: the pollution of the ocean, which has assumed dangerous proportions, and the impoverishment of its biological resources are by no means irreversible. These disasters can be averted. If all of us on earth pool our efforts to keep the ocean clean and sea-life intact, the ocean will amply reward us.

The mysteries and excitement of the underwater world seem far away as the sun sets over these tranquil waters.

The old castle on a cliff which houses the world-famous oceanographic museum at Monaco, headquarters of Jacques-Yves Cousteau.

ACKNOWLEDGMENTS

APN (Novosti Press Agency): pp. 11, 13 (top), 20 (bottom), 66, 76, 78 (top), 81, 87, 92 (bottom left), 100, 101 (top), 102 (top), 108 (top). 109. R. Catala: pp. 30 (bottom), 31 (top right), 37 (top), 67 (top), 83 (left), 86 (left), 105 (second top). CNEXO: pp. 10 (bottom), 12 (bottom left), 15 (both), 24 (top), 54, 104, 114 (bottom). COMEX: pp. 13 (bottom), 60, 64–5, 88, 90 (top), 92–3 (top), 95. J. Y. Cousteau: pp. 4 (second top), 42, 46, 47, 49, 50, 51 (bottom), 53 (top), 56–7, 70 (right), 101 (top). Cousteau Society/FPG: pp. 35. N. Denisov: pp. 78 (bottom). FPG: pp. 2, 6, 10 (top), 14, 20 (second bottom), 26, 27, 28, 30 (top left), 30–1 (top), 38, 39, 44, 45, 46, 52 (bottom and second top), 55 (bottom), 58, 61 (both), 62 (bottom and second bottom), 68, 71, 72, 73 (top), 80 (top), 82, 83 (right), 85 (both), 94 (bottom and second bottom), 103 (bottom), 105 (top), 111, 118 (top), 123, 124 (top). Globe Photos, New York: pp. 21, 59 (Don Dornan), 120. V. Kasho: pp. 10 (second top), 29, 37 (bottom), 40, 41, 52 (second bottom), 55 (top), 62 (top and second top), 67 (bottom and second bottom), 70 (left), 73 (bottom and second bottom), 74 (all), 80 (second bottom), 94 (top), 102 (bottom), 118 (second top), 122 (second bottom), 124 (left), 125 (top). Yu. Muravin: pp. 122 (top). S. Rybakov: pp. 51 (top). P. Spirkov: pp. 4 (second bottom), 20 (top), 63 (bottom), 79 (bottom), 91 (bottom), 93 (bottom), 97, 110. V. Suetin: pp. 63 (top), 80 (bottom). US NAVY/FPG: pp. 24 (bottom), 107, 114 (top), 115. G. Ventouillac: pp. 10 (second bottom), 52 (top), 80 (second top), 118 (second bottom), 121 (top and second top), 122 (bottom), 125 (bottom), V. Yuksha: pp. 75. Authors' archives: pp. 8, 9, 25 (all), 34, 53 (bottom right), 79 (top), 90 (bottom), 94 (second top), 96, 98, 103 (top), 112, 118 (bottom).

Drawings by de Neville: pp. 4 (top), 32. V. Radayev: pp. 12 (both top), 36, 48–9, 77, 87, 105, 106, 108–9, 116, 116–7, 119. R. Varshamov: pp. 22–3. Maps by V. Khramov: pp. 16–7, 18–9.

Cover: Elgin Ciampi/Freelance Photographers Guild (FPG), New York. Front endpapers: Ron Church/FPG. Back endpaper: Jack McKenney/FPG.

FURTHER READING

Projects and Pioneers
Chapman, Sydney. *I.G.Y., Year of Discovery: The Story of the International Geophysical Year.* Ann Arbor, Michigan: University of Michigan Press, 1959.

Cousteau, Jacques. *Outer and Inner Space.* Ocean World of Jacques Cousteau, vol. 15. New York: Harry N. Abrams, 1975. London: Angus and Robertson, 1976.

————. *Window on the Sea.* Ocean World of Jacques Cousteau. New York: Harry N. Abrams, 1973. London: Angus and Robertson, 1974.

*Daugherty, Charles M. *Searchers of the Sea: Pioneers in Oceanography.* New York: Viking Press, 1961.

*Dugan, James. *Underwater Explorer: Story of Captain Cousteau.* New York: Harper & Row, 1957.

Evans, Idrisyn O. *Observer's Book of the Sea & Seashore.* Observer's Pocket Series. New York: Warne, 1962. London: Warne, 1962.

Heintze, Carl. *The Bottom of the Sea and Beyond.* New York: Thomas Nelson, 1975.

Idyll, C. P., ed. *Exploring the Ocean World: A History of Oceanography.* rev. ed. New York: T. Y. Crowell, 1972.

*Pennington, Howard. *The New Ocean Explorers: Into the Sea in the Space Age.* Boston: Little, Brown, 1972.

General
Borgese, Elizabeth M. *Drama of the Oceans.* New York: Harry N. Abrams, 1976.

Brown, Seyom et al. *Regimes for the Ocean, Outer Space and Weather.* Washington, D.C.: Brookings Institution, 1977.

*Carson, Rachel. *The Sea Around Us.* rev. ed. New York: Oxford University Press, 1961.

Groen, Pier. *Waters of the Sea.* New York: Van Nostrand Reinhold, 1967.

*Hyde, Margaret O. *Exploring Earth and Space.* 5th ed. New York: McGraw-Hill, 1970.

Piccard, Jean. *Sun Beneath the Sea.* New York: International Publications Service, 1973.

Soule, Gardner. *Greatest Depths: Probing the Sea Below Twenty Thousand Feet.* Philadelphia, Pa.: Macrae Smith, 1970.

Vinogradov, A. P. and Udentsev, G. B., eds. *The Rift Zones of the World Oceans,* trans. N. Kaner. New York: Halsted Press, 1975.

Wegener, Alfred. *Origin of Continents and Oceans,* trans. John Biram. New York: Dover, 1966. London: Methuen, 1968.

Wilson, J. Tuzo, ed. *Continents Adrift and Continents Aground.* Readings from Scientific American. San Francisco: W. H. Freeman, 1976.

Special Aspects
Anderson, Frank J. *Submarines, Diving and the Underwater World: A Bibliography.* Hamden, Ct.: Shoe String Press, 1975.

Caruthers, J. W. *Fundamentals of Marine Acoustics.* Elsevier Oceanography Series, no. 18. New York: Elsevier, 1977.

Clay, Clarence S. and Medwin, Herman. *Acoustical Oceanography: Principles and Applications.* New York: John Wiley, 1977.

Geyer, R. H. *Submersibles and Their Use in Oceanography and Ocean Engineering.* Elsevier Ocean Series, no. 17. New York: Elsevier, 1977.

*McFall, Christie. *Underwater Continent: The Continental Shelves.* New York: Dodd Mead, 1975.

McQuillen, R. *Exploring the Geology of Shelf Seas.* New York: International Publications Service, 1975.

Penzias, Walter and Goodman, M. W. *Man Beneath the Sea: A Review of Underwater Ocean Engineering Technology.* New York: John Wiley, 1973. London: Interscience, 1973.

Shenton, Edward L. *Diving for Science: The Story of the Deep Submersible.* New York: Norton, 1972.

Shepard, Francis P. and Dill, Robert F. *Submarine Canyons and Other Sea Valleys.* New York: John Wiley, 1966. London: John Wiley, 1971.

Woods, J. D. and Lithgoe, J. N., eds. *Underwater Science: An Introduction to Experiments by Divers.* New York: Oxford University Press, 1971.

Policy for the Future
Doumani, George. *Ocean Wealth: Policy and Potential.* Rochelle Park, N.J.: Hayden Book Co., 1973.

Friedman, Walter. *The Future of the Oceans.* New York: George Braziller, 1971.

Jada, Lawrence. *Ocean Space Rights: Developing U.S. Policy.* New York: Praeger, 1975.

*Ross, Frank, Jr. *Undersea Vehicles and Habitats: The Peaceful Uses of the Ocean.* New York: T. Y. Crowell, 1970.

*For secondary school students.